EXTINCT

The Complete Visual Guide to Extinct and Endangered Species

General Editor	Tamara Green
Designer	Marilyn Franks
Cover Designer	Emily Gibson
Illustrators	Tony Gibbons
	Helen Jones
	Neil Lloyd
Consultant	Dr Paul Barrett

First published in Great Britain in 2001 by Brimax
an imprint of Octopus Publishing Group Ltd
2-4 Heron Quays, London E14 4JP

ISBN 1 8585 4398 3

Printed in China

A CIP catalogue record for this book is available
from the British Library.

EXTINCT

The Complete Visual Guide to Extinct and Endangered Species

BRIMAX

Contents

Introduction

As you turn the pages of this splendidly illustrated encyclopedia, you will find yourself embarking on a most unusual sort of safari – quite simply because all the animals presented are either extinct or terribly rare.

Dinosaurs, of course, disappeared 65 million years ago; and creatures such as the mammoths and sabre-tooth tigers have also long since vanished. But many animals have died out far more recently – the quagga, for instance, the dodo, and the giant beaver. And with every year that passes, more and more, large and small, seem destined to disappear from our world, never to be seen again.

We invite you now to meet many forms of wildlife that, alas, no longer grace Planet Earth.

Discover, as you read on, why extinction has so often occurred. All-importantly, too, find out what might be done in an attempt to save the many endangered species, like the panda *below*, without which the world would be a far poorer place.

In prehistoric times

Some of the creatures that roamed Planet Earth many millions of years ago are among the strangest you can imagine. Not only were there dinosaurs, but mighty beasts such as Glyptodon, Brontotherium, and the giant ground sloth.

But did you know that there was a dinosaur that, as well as being a meat-eater, also enjoyed meals of fish? At that time there were also huge monsters, larger than some of today's light aircraft, that flew in the air, and sea creatures that could swim at speeds of up to 25 miles (40km) per hour.

MORE AND MORE CURIOUS

Some of the animals that evolved after dinosaurs became extinct were even odder. One, for instance, had such a large shell that, long after it disappeared from Planet Earth, human beings used this part of its body to make shelters for themselves.

Way back in time there were also giant marsupials that carried their young in pouches, just like the much smaller kangaroos and opossums do today. There were huge flightless birds, too, and extraordinary winged lizards.

Every year, exciting new discoveries are made. More and more fossils of previously unknown forms of prehistoric life are unearthed, and scientists work tirelessly to build up a picture of what they must have looked like when alive. Sometimes there are very exciting surprises in store.

As you will discover, for example, a species thought to have become extinct many millions of years ago was found to have survived into modern times by fishermen trawling in the Indian Ocean. Read on to find out all about this highly unexpected discovery!

And what of the creatures on these two pages? Perhaps you cannot identify each of them now. But by the time you have completed this most unusual type of safari, you will be an expert in your own right on prehistoric life. Which of them will you find most fascinating?

Pareiasaurus

Approximately 250,000,000 years after she walked alive on Planet Earth back in Permian times, a mighty Pareiasaurus, nicknamed Delilah, has been unearthed in South Africa.

In 1998, part of the 6.5ft (2m)-long skeleton of this early reptile became exposed on a mountainside. It was found in a crouching position, and its many interlocking plates were still covering its back. As possibly the best preserved specimen of this species, it was an incredible find. However there was one major problem. How could the team excavate it speedily while keeping it intact?

LIVED: in Permian times
SIZE: 6.5ft (2m) long
WEIGHT: 1,543 lbs (700kg)
DISCOVERED: in South Africa, in 1998

OTHER DATA: very bumpy protective covering; a powerful frame; stocky legs; possibly semi-aquatic; skull of a young specimen; *say:* PA-REYE-OH-SAW-RUS; status: extinct

First they dug a trench around Delilah, and then they encased her in plaster of Paris to protect her skeletal remains. Unfortunately, however, no craft could be found to provide an adequate airlift.

There was nothing else for it. The team, assisted by a score of tough young men from nearby farms, had to combine forces and lift what they estimated to be a 1,543 pound (700kg) creature. After much huffing and puffing, they made it, so that eventually the Pareiasaurus, could be put on public display.

A FITTING NAME?

Delilah, as you might have guessed, was given her name with tongue in cheek. After all, she was hardly as beautiful as her namesake, who betrayed Samson in the well known Bible story! Indeed, reconstructions, like the one in the illustration *left*, show Pareiasaurus to have had a formidable appearance, its ugly, bulky body covered with a knobbly outer coating formed from bony protuberances. Its head, too, was protected in this way.

In fact, its eyes were hardly visible among all those bumps of bone, and seemed embedded in the uneven surface. It also had very broad limbs and stocky legs which were needed to support its powerful frame.

UNTIMELY END

Living long before the dinosaurs had evolved, Pareiasaurus met its demise during the mass extinction that seems to have occurred at the end of Permian times, about 250 million years ago. It has even been estimated that about 95 per cent of all species existing then completely disappeared.

No one is entirely sure why this happened. Some experts think the climate may have changed dramatically so that it could no longer support life as it was then. Others contend there was a marked change in sea level; that the lava and ash from extensive volcanic eruptions may have devastated both flora and fauna; or that an asteroid hit Earth. Whatever the cause, only a few species survived. Pareiasaurus does not seem to have been one of them.

World of Pareiasaurus

- The skeleton of a young Pareiasaurus, complete with skull, was found in the region of Kotelnich, Russia.
- Some scientists believe that Pareiasaurus may have been semi-aquatic, and that it possibly evolved into turtles.

Dimetrodon

Among the most advanced creatures of its time, Dimetrodon was longer than today's average-sized car; and, from the size and shape of its teeth, scientists can tell it must have been a fearsome carnivore.

One of the most successful predators of its era, Dimetrodon had a spiny "sail" on its back that gave it a particularly menacing look. This structure probably also played a major part in Dimetrodon's success as a hunter.

The sail, we think, provided a kind of personal temperature-control system. Dimetrodon is thought to have been able to use the sail either to heat itself up or to cool down. Readily warming up on cool mornings, Dimetrodon was thus always one vital step ahead of its far more sluggish potential victims.

World of Dimetrodon

- Dimetrodon lived in Early Permian times, about 280 million years before humans first evolved.

LIVED: in Early Permian times
SIZE: 11ft (3.3m)
WEIGHT: unknown, but heavy
DISCOVERED: in North America

OTHER DATA: structure like a sail on its back, probably useful for temperature control; long tail; chunky legs; carnivore; strong bite; *say:* DYE-MET-ROH-DON; status: extinct

Dimetrodon's "sail" may also have had another function, too. It could have been a form of sexual display, males perhaps having a larger and brighter sail than the females.

Dimetrodon certainly looked something like a dinosaur, but it was not one. A pelycosaur, it died out even before dinosaurs first evolved.

Precisely why Dimetrodon became extinct when it did is something of a mystery, but some scientists believe it may have met with considerable competition from other more advanced creatures.

Lagosuchus

The creatures seen *below* date from Triassic times and are so similar in appearance to some of the dinosaurs that some scientists think they may have been a sort of missing link from which dinosaurs such as Staurikosaurus, one of the very first, evolved.

The two Lagosuchus in this illustration, seen scampering around on a rocky outcrop, were carnivores although only the size of rabbits. Indeed, their name actually means "leaping crocodile." They were given this name by the American palaeontologist Alfred Sherwood Romer (1894-1973).

SMALL AND SPEEDY

Since it was so small, Lagosuchus could bound around very easily and would dig into crevices with its snout to find insects and bugs of all kinds on which to feast. When Lagosuchus ran along, it probably used its back legs only, but sometimes it would have gone about on all-fours. It might also have been able to spring out at small animals, using its long back legs to propel it forward at speed.

LIVED: in Triassic times
SIZE: rabbit-sized
WEIGHT: lightweight
DISCOVERED: in Argentina

OTHER DATA: legs like a rabbit's; hands capable of grasping prey; other features resembling those of dinosaurs; long, slim tail; speedy; *say:* LAR-GOH-SOOK-US; status: extinct

They needed to be fast on their feet and very wary. If they did not have all all their wits about them, a group of several Lagosuchus might easily have risked becoming a meal for carnivorous predators.

Lagosuchus may even have been agile enough to climb trees to avoid an enemy, and might have ventured into shallow water at times, although it had no features that suggest it was a strong swimmer.

> **FACTFILE**
> Lagosuchus remains, dug up in Argentina, South America, date from just before dinosaurs first evolved 230 million years ago.

DINOSAUR-LIKE

Which particular features, then, have made many palaeontologists believe that Lagosuchus may well have been a close relative or even a direct ancestor of the dinosaurs?

First of all, it could stand and run on its two back legs. Its knee joints could be straightened, and the lower part of the leg was longer than the upper part It also ran on its toes, rather than using the soles of its feet. Lagosuchus could stand erect to survey its surroundings, too.

ON BALANCE

Such an upright gait is even believed to have been one of the secrets of the success of several species of dinosaur.

Lagosuchus also held its legs beneath the body for much of the time; and just like many of the dinosaurs that were soon to evolve, it could maintain its balance very effectively when running at speed with the help of its long, slender tail.

Interestingly, one of the oldest known and most primitive of the dinosaurs, Staurikosaurus, was also first unearthed in South America. It was considerably larger than Lagosuchus, however, measuring about 6.5ft (2m) in length, and weighed about 66 pounds (30kg).

On pages 16-39, you can find out what true dinosaurs were like, when they evolved, the type of landscape in which they lived, and how long they existed on the globe before dying out.

World of Lagosuchus

- In Triassic times, when Lagosuchus existed, the climate was warm and there were heavy rains, but not in inland regions where vast deserts formed.
- Ginkgoes, conifers, cycads, horsetails and ferns were the main vegetation, and all the land was fused together into one super-continent, known as Pangaea.

Triassochelys

Looking something like a turtle or the tortoise of today but not yet able to withdraw its head to hide from the world, Triassochelys had a shell far wider than it was long.

From its name, it is easy to guess that Triassochelys must have lived in Triassic times.(Both words start with the same first six letters.) The shell of this primitive testudinate provided very strong body protection for its far softer inner body, and it was about 3ft (0.9m) in length.

Remains have been found in central Europe, and a well preserved fossil, found near Trossingen in Germany, can be seen at the Natural History Museum in Berlin.

The large shell covering gave the appearance of being richly sculpted. Triassochelys also had a neck and tail that were protected to some degree by small spikes. However, these would probably not have been sufficiently menacing to deter any highly determined, hungry Triassic predator from attacking a Triassochelys.

LIVED: in Triassic times
SIZE: 3ft (0.9m) long
WEIGHT: unknown
DISCOVERED: in Germany and elsewhere

OTHER DATA: early turtle; probably could not withdraw head and tail when in danger as later turtles can; no teeth; a herbivore; beak; *say*: TRY-ASS-OK-EEL-ISS; status: extinct

From its remains, scientists can tell that Triassochelys lived mostly on land. We know, too, that it had no teeth and that it was an herbivore, using its primitive beak to bite off vegetation.

The largest testudinate ever to have existed did not evolve until about 80,000,000 years ago, in Cretaceous times. Known as Archelon, it grew to 13ft (4m) in length and may have weighed as much as 3 tons. Unlike Triassochelys, it lived only in the water.

World of Triassochelys

- Remains of this testudinate can be seen in the British Museum of Natural History, London, as well as in Berlin.
- Scientists think it was not only its head that Triassochelys could not withdraw into its shell when faced with a predator, but its tail, too.

Dinosaurs

Not all dinosaurs lived at the same time during the 160 million years for which they ruled the world, long before human beings existed. Some died out and others then evolved. But they walked this planet for far longer than humankind has done so far – 50 times as long, in fact.

There were many different types of dinosaur. Some were the most fearsome meat-eating monsters, while others were far more docile creatures that lived in fear of being attacked by predatory beasts such as Cretaceous Daspletosaurus. This dinosaur was in fact a member of the same group of dinosaurs to which Tyrannosaurus rex belonged. As you can see in this illustration, even those herbivores that lived in large herds would run for their lives when this mighty beast was on the prowl.

Dinosaurs first evolved in Triassic times which lasted from approximately 248-213 million years ago. Then came another period of prehistory, the Jurassic era, which spanned 213-144 million years ago. This was followed by Cretaceous times, at the end of which dinosaurs finally became extinct 65 million years ago. The next few pages are devoted to some of the most intriguingly different dinosaur species discovered so far.

FACTFILE

Dinosaurs were not all huge carnivores like Tyrannosaurus rex. Some only ate plants, while others were the size of a chicken.

LIVED: in Triassic, Jurassic and Cretaceous times
SIZE: from that of a chicken to giant proportions
WEIGHT: from a few pounds to many tons
DISCOVERED: worldwide

OTHER DATA: some were herbivores, some were carnivores; a few ate fish; could not swim; fossilized tracks in places; some lived in herds; tough skin; status: extinct

World of dinosaurs

- The word *dinosaur* means "terrible lizard" and was given to this group of land animals by a great British palaeontologist, Sir Richard Owen.

Ankylosaurus

Two Ankylosaurus, their mighty tail clubs raised high in the air, stand ready for battle. They have heard the roar of a mighty predator in the distance and may need to use their body armour in self-defence.

These huge and stocky plant-eaters lived in what is now Alberta, Canada, and Montana in the United States during Cretaceous times. All those body spikes and the dreaded tail weapon provided excellent protection from such predators as Tyrannosaurus rex. Indeed Ankylosaurus was so well armoured that it has even been described as a living tank.

When defending itself, it probably stood with its back to its enemy and then gave that heavy tail club a terrific swing. This would have dealt a dreadful blow, stunning the attacker to such an extent that it may have dropped to the ground, then lying there in a stupor.

SOFT-BELLIED

Its only point of weakness would have been its soft belly, which was why an enemy would have gone all out to turn Ankylosaurus on to its back so that it could take a bite at its flesh and bleed the creature to a slow death.

With a name meaning "jointed lizard," given because of the bony plates that were fused to each other and embedded into its skin, Ankylosaurus is thought to have had the toughest sort of body covering of any creature ever known to have existed.

It stood about 8ft (2.5m) tall and was up to 33ft (10m) in length; and as an extremely stocky creature, weighed at least 2 tons.

LIVED: in Late Cretaceous times
SIZE: 33ft (10m) long
WEIGHT: at least 2 tons
DISCOVERED: in 1908, in Canada

OTHER DATA: quadruped; beaked mouth; herbivore; clubbed tail; soft belly and underparts; name means "fused reptile;" sturdy legs; *say:* AN-KILL-OH-SAW-RUS; status: extinct

Ankylosaurus was one of the last dinosaurs to live in what is now North America, alongside ceratopsians such as Triceratops and tyrannosaurids such as the monster Tyrannosaurus rex. Then, 65,000,000 years ago, they all became extinct.

World of Ankylosaurus

- Remains of Ankylosaurus have been found in the state of Alberta, Canada, as well as in Montana in the United States.

Maiasaura

When palaeontologists Jack Horner and Robert Makela visited a fossil shop in Montana in 1978, they found the bones of what seemed to be a baby hadrosaur and rushed to where they had been unearthed.

In all, Horner, Nakela and their team found fourteen nests, thirty-one babies and forty-two eggs at the site. But that was only the start of it. Some years later, in 1984, Horner discovered a bone bed where as many as ten thousand Maiasaura had perished. One theory is that they died there as the result of a volcanic eruption, but we may never know for sure.

This 30ft (9m)-long Cretaceous dinosaur made bowl-shaped nests from mud and sand, and would line them with soft vegetation. The nests would be made in colonies, and the mother Maiasaura would curl up around her eggs at night to keep them warm. Each nest was about 3-4ft (0.9-1.2m) deep and 6.5ft (2m) in diameter. The babies would have been just 14ins (35cm) long when they hatched, but they doubled in size within two months.

FACTFILE

The name *Maiasaura* means "good mother lizard" and is most suitable for a dinosaur unearthed with lots of eggs and babies.

LIVED: in Late Cretaceous times
SIZE: 30ft (9m) long
WEIGHT: 6 tons
DISCOVERED: in North America, in 1978

OTHER DATA: herbivore; name means "good mother lizard;" a hadrosaur; lived in herds; mostly went on all-fours; *say*: MY-A-SAW-RA; status: extinct

World of Maiasaura

- From the Maiasaura nesting site, we know that dinosaurs did not abandon their young after they had hatched but continued to care for them until they were able to look after themselves.

Strictly an herbivore, and feeding mostly on twigs and leaves for which its many grinding teeth were ideal, Maiasaura is thought by some scientists to have fed its young by first chewing on the vegetation and then regurgitating it for them, so that they did not have to cope with tough plants.

GOING WITH THE HERD

Like most hadrosaurs, Maiasaura lived in large herds which migrated from time to time in search of new feeding grounds. Palaeontologists have been able to tell this because of the extensive tracks they have found. Predators may have threatened them at intervals, but there would have been safety in numbers for these large, docile dinosaurs.

Stegosaurus

A dinosaur from Jurassic times, Stegosaurus had only a walnut-sized brain; yet through its body plates it had both an ingenious way of controlling its temperature and defending itself.

Most of the time, Stegosaurus had to watch out for many types of greedy predator. But whenever it could, this plated dinosaur would use its toothless beak to chomp on any available vegetation.

LIVED: Late Jurassic times
SIZE: 30ft (9m)
WEIGHT: 6-8 tons
DISCOVERED: in North America, in 1877

OTHER DATA: herbivore; small head; toothless beak; large plates on its back, possibly to control temperature; small cheek teeth; *say*: STEG-OH-SAW-RUS; status: extinct

The skeletal remains of Stegosaurus are easily identifiable because of the magnificent array of plates that ran all along this creature's back. These plates provided superb body-armour that would protect it at least to some degree from predators such as greedy Allosaurus. Its tail, meanwhile, was spiked and so could be used to give an effective sideways swipe.

But even when no enemy was at hand, those enormous plates are thought to have been put to good use. If Stegosaurus turned sideways at the start of a Jurassic day, when the temperature was probably fairly cool, so that the flat sides of the plates faced the sun, its whole body would gradually warm up

Later in the day, however, it may have been too hot. Then it could turn so the plates would not absorb so much heat.

In many ways, such a heating system is similar to the heat-absorbing solar panels which some very modern buildings have today. Scientists have even found markings on the plates that seem to show there were blood vessels within the plates. At first, some even suggested that these plates may have been adjustable so that they could be made to lie flat when not in the usual upright position. But this theory is not now widely accepted.

World of Stegosaurus

- Jurassic Stegosaurus lived in regions now forming the states of Colorado, Oklahoma, Utah and Wyoming, USA.

Mamenchisaurus

A dinosaur record-breaker and the longest of any animals known to have existed, Mamenchisaurus was first unearthed in China. Its neck alone extended for 49ft (15m), and this was balanced by an exceedingly long and tapering whiplash tail.

The longest of all the Jurassic sauropods, Mamenchisaurus must have given the impression of being a very proud creature because palaeontologists think that it held its head upright for most of the time. So just imagine how far it would have had to swing its neck down when it wanted to drink from a pool!

Each of the 2ft (60cm)-long neck bones must have been very lightweight, however, or Mamenchisaurus could never have lifted its head.

SPECIAL TEETH

Given its name in 1954 by the great Chinese palaeontologist Yang Zhong-jian, Mamenchisaurus was different from many of the other sauropods. Indeed, those sauropods found in China so far all have spatulate teeth. This means that their teeth were far wider and thicker than the peglike teeth of sauropods such as Diplodocus, found in the United States.

LIVED: in Jurassic times
SIZE: up to 89ft (27m) long
WEIGHT: 25 tons
DISCOVERED: in China, in 1954

OTHER DATA: a sauropod; herbivore; longest known neck of any dinosaur; walked on four legs; tiny head in comparison with body; *say*: MAM-EN-KEE-SAW-RUS; status: extinct

Mamenchisaurus would certainly have had no competition when browsing in the treetops because, quite simply, no other dinosaur would have been able to reach up so high.

Another sauropod with a terrifically long neck was the Jurassic dinosaur Giraffatian, found in Tanzania, Africa. But it came a poor second, even though its name means "gigantic giraffe." Its neck was only about two-thirds the length of the neck of Mamenchisaurus and extended to only 33ft (10m).

Mamenchisaurus probably weighed about 25 tons, so it was by no means the heaviest of the dinosaurs in spite of its tremendously lengthy neck. Seismosaurus weighed over 100 tons!

FACTFILE

Mamenchisaurus was named after a watercourse in the region of China near which its remains were first found.

World of Mamenchisaurus

- When the bones of Mamenchisaurus were first unearthed, there were no teeth and no complete skull with the remains. Since then, however, many more fossils of Mamenchisaurus have been found in China, including an almost complete skull.

Deinonychus

A prime example of dinosaurs that worked in packs to overcome a larger victim and then fed together on the carcass, Deinonychus had deadly switchblade weapons on each of its second toes.

In the illustration shown here, two ferocious Deinonychus are about to tear a terrified Tenontosaurus to pieces with their powerful claws and sharp teeth. It would have taken them very little time to overcome this large dinosaur.No wonder it was given a name meaning "terrible claw."

First dug up in a hillside in the state of Montana, North America, by palaeontologist John Ostrom and his team in 1964, Cretaceous 10ft (3m)-long, sharply clawed Deinonychus was clearly an effective predatory pack animal.

World of Deinonychus

- Deinonychus had lots of backward-curving teeth which had serrated edges, just like today's steak knives. Once it had gripped a victim with these very sharp teeth, there was no pulling away from its hold.

Deinonychus's two switchblade claws grew to 5ins (13cm) in length, but they did not prove a hazard to this dinosaur when it was walking or running at speed. Then, Deinonychus would use powerful muscles to hold the toes which held the switchblade claws clear of the ground, so there was no risk it would trip up and the outsize claw did not become blunt by trailingon the ground.

LIVED: Early Cretaceous times
SIZE: 10ft (3m) long
WEIGHT: 155 pounds (70kg)
DISCOVERED: in North America, in 1964

OTHER DATA: carnivore; large clawed front limbs; sickle claw on each back limb; long tail; hunted in packs; sharp teeth; powerful bite; *say:* DEYE- NOH-NEYE-KUS; status: extinct

When attacking, Deinonychus would kick at its quarry with a hind leg. The switchblade claw would then flick forwards forcefully and dig into the victim's flesh. As it did this, its stiff bony tail would be held aloft, helping it to balance. This sort of tail was certainly a useful body tool. It could be swished from side to side as Deinonychus ran, enabling the dinosaur to dodge about with ease as it tried to catch up with its prey and then corner it.

Deinonychus also had sharp, curved finger claws that would have been used to dig into a victim's flesh, causing a terrible wound and an excruciatingly painful reaction.

For its body size, Deinonychus must have had a large brain and so was probably rather wily.

It would have been able to learn from experience, some palaeontologists have suggested, and so remembered the best tactics for attack.Prior to the discovery of Deinonychus, the prevalent theory was that dinosaurs had been quite stupid, slow-moving creatures, but we now know differently.

FACTFILE

Deinonychus's skull was large but very lightweight because there were holes, known as "windows," within it.

Iguanodon

Perhaps you usually think of palaeontologists and fossil-hunters as men. Many women, however, have become experts in prehistoric remains, and not only recently. It was, in fact, a British woman, Mary Ann Mantell, who found, by complete accident, a tooth from a remarkable dinosaur which her husband, Dr Gideon Mantell, then named Iguanodon three years later in 1825.

FACTFILE

Remains of Iguanodon, whose name means "iguana tooth," have also been unearthed in the USA and in Mongolia.

The first fossilized part of Iguanodon to be found was indeed that tooth. But it is actually for another very distinctive feature that Iguanodon has become known – the large spiked thumb on each of its forelimbs.

Palaeontologists have discovered that, for most of the time, Iguanodon would have been a peaceful herbivore, spending its day simply browsing on vegetation or scampering about along with the rest of the herd, as in this illustration. But when a predator threatened, Iguanodon would have reared up on its hind legs and used its spiked thumbs to dig into the neck of the attacker. A single Iguanodon probably saved its life many times with this built-in weapon in this way.

Amusingly, when scientists first examined Iguanodon's thumb spike, they thought it must have belonged on its nose; but now, of course, we know it was attached to a thumb!

Iguanodon had a head shaped much like that of a horse and a toothless horny beak which was ideal for biting off leaves. But it did have lots of cheek teeth which it used to grind up vegetation.

Some of the skeletal remains of Iguanodon are larger than others, and this has led some palaeontologists to speculate that the males and females may have varied in size. But whether the males were larger is not known for sure. In modern reptiles, it is sometimes the females that are bigger.

World of Iguanodon

- Iguanodon lived in the Cretaceous period, when flowering plants began to appear for the very first time.
- When Iguanodon walked our planet, earthquakes and volcanoes were far more common than they are today.

LIVED: in Early Cretaceous times
SIZE: 30ft (9m) long
WEIGHT: 5 tons
DISCOVERED: in England, in 1822

OTHER DATA: herbivore; heavy tail; spiked thumb weapons used to defend itself; lived in herds; self-sharpening beak; cheek teeth; *say*: IG-WAN-OH-DON; status: extinct

Coelophysis

In 1947, an entire dinosaur graveyard was unearthed at a site in New Mexico, known as Ghost Ranch. It was an amazing discovery. Not only were the remains of over one hundred very early dinosaurs unearthed, some of them turned out to have tiny fossilized baby skeletons in their stomach cavities.

Some dinosaurs were herbivores, eating only vegetation and not hunting for prey. Others, however, were greedy meat-eaters and therefore highly predatory. But scientists now know, too, that at least some were cannibalistic.

Since dinosaurs laid eggs and did not give birth to live young, the tiny skeletons found in the stomach cavities of the Coelophysis specimens at Ghost Ranch must have belonged to babies that had been eaten. So it seems that Coelophysis – and perhaps other dinosaurs, too – resorted to cannibalism when food was in short supply. Maybe the flesh of their young was even regularly on the menu. They may also have raided the nests of others of their species whenever none of their own offspring was available.

Coelophysis (the name means "hollow form") had a slim body and was only 10ft (3m) in length from the tip of its snout to the end of its tail. It was not heavy, weighing only about 65 pounds (30kg).

FACTFILE

The Triassic dinosaur Coelophysis was first described by the eminent palaeontologist Edward Drinker Cope in the year 1889.

LIVED: Late Triassic times
SIZE: 10ft (3m) long
WEIGHT: 44lbs (20kg)
DISCOVERED: in North America, in 1881

OTHER DATA: carnivore; probably hunted in packs; cannibalistic, even eating its own young; long, narrow jaws; intelligent; *say:* SEEL-OH-FEYE-SIS; status: extinct

But, in spite of this, Coelophysis must have been a feared predator because of its very sharp teeth. It is also likely to have lived and hunted in packs.

Most large carnivorous dinosaurs existed in either Jurassic or Cretaceous times. Those of the earlier Triassic period – Coelophysis, for example – were far smaller; and some were only the size of today's chickens.

World of Coelophysis

- Coelophysis was among the very first dinosaurs to have evolved back in the Triassic. Few smaller dinosaur fossils have been dug up because they disintegrated with time.

Did sauropods swim?

One of the dinosaurs that scientists once thought could swim was Brachiosaurus, a sauropod. As tall as a four-storey house and with a head featuring high-placed nostrils, 250 tons of bones from these enormous Jurassic herbivores were first discovered in Tanzania, Africa.

World of dinosaurs

- The long tails of the sauropods would have helped them to keep their balance when walking in the water.

For a number of years after it had first been unearthed, scientists thought that Brachiosaurus and other sauropods like the one in the illustration, *below,* might have been able to swim. They believed that some dinosaurs may even have lived in lakes and rivers rather than on land. The water, these scientists said, would have helped to support their huge bodies as they raised their long necks above the water; and as herbivores, they could have fed on marine plant life.

Now, however, palaeontologists think quite differently and hold that dinosaurs spent most of their time on land.

LIVED: most sauropods lived in Jurassic times
SIZE: most were very large
WEIGHT: many tons
DISCOVERED: in North America, China, Germany etc

OTHER DATA: sauropods could not swim but trod water in the shallows, as their fossilized footprints show; probably went into the water to cool down; status: extinct

Only occasionally did dinosaurs venture into the water.

There are several reasons for this change of opinion among the experts. Firstly, it would have been impossible for such a creature to withstand the water pressure and to breathe properly at any great depth, they explain.

A typical sauropod would not have swum in the true sense of the word but, when venturing into lakes, rivers or the sea.

Instead, it would "tread" water, walking along in the shallows. Here it may have fed on waterplants.

Going into the water may also have helped to cool the sauropods down on very hot Jurassic days when temperatures would have soared to heights far greater than those we know today. It might also have enabled them to avoid attack. Some sauropods were so tall, remember, that they could no doubt go quite far into the water before risking getting drowned.

FACTFILE

If dinosaur remains are found in the sea, this does not mean they swam but that waters now cover the land they once inhabited.

Baryonyx

Some dinosaurs were carnivores and ate mostly meat, killing other species for food. Others were strictly herbivores, keeping to a diet of vegetation only. At least one species, however, liked fish.

Proof that Cretaceous Baryonyx liked fish, in addition to meat, comes from the fact that some undigested fish scales were found in the region of the rib cage of its skeletal remains. Remarkably, scientists have even been able to identify the type of fish from which these scales came. It was a 3ft (0.9m) long prehistoric Lepidotes, which was an ancient relative of today's sturgeon.

Undoubtedly the most distinctive feature of Baryonyx was the 12-inch (30cm)-long thumb claw on each forelimb. These claws seem to have been extremely sharp and would have been excellent weapons to use against predators. They would also no doubt have served as useful tools for stabbing at any slippery marine victims or land-based, plant-eating prey, although Baryonyx's long jaws, as you can see in this illustration, were ideally suited to snapping these up, too.

FACTFILE

Baryonyx probably weighed as much as 2 tons, and scientists think it may have got about on all four limbs or on two at times.

AMATEUR FIND

The very first remains of Baryonyx to be found were unearthed by an amateur fossil hunter called William Walker in 1983. He felt it was likely he would discover dinosaur bones in a particular pit because certain specimens had already been found there, and he was proved right. What he found was a massive claw that had broken into several pieces.

His next step was to take the remains to the Museum of Natural History in London, where palaeontologists found the claw of such interest that they arranged for a professional dig to take place at the pit.As as result much of the skeleton and skull, nearly 3ft (0.9m) long, were also dug up.

The head, it was found, was very much like that of a crocodile, and belonged to a creature that was in all about 32ft (10m) long. Like a crocodile, Baryonyx may even have swum in pursuit of its prey. It has even been said that, when at the water's edge, it might have grappled with fish using its claws in much the same way as a bear will try to pull a salmon from a river. It is interesting that Baryonyx remains were found alongside those of Iguanodon, another dinosaur with a spiked thumb weapon. Perhaps they fought with these.

LIVED: in Early Cretaceous times
SIZE: 30ft (9m)
WEIGHT: 2 tons
DISCOVERED: in England, in 1983

OTHER DATA: carnivore; ate meat and fish, caught with thumb claws; crocodile-like jaws; some of best preserved dinosaur remains are of Baryonyx; *say*: BA-REE-ON-IKS; status: extinct

Sauropods

The biggest of the sauropods were the largest creatures ever to exist on Earth. But though more than twice the size of contemporaries such as Allosaurus, for instance, they were not fierce.

Easily recognisable by their long necks, thick pillarlike limbs, small heads and lengthy, tapering tails, sauropods – such as Diplodocus, Brachiosaurus, Apatasaurus, and Cetiosaurus – were gentle giants that lived in huge herds of their own kind.

They ate almost constantly and would swallow all the vegetation that they consumed without chewing it. Their digestion did not suffer, however, because they would also take in stones, known as gastroliths, which would grind up all the plantstuff in their stomachs.

As they wandered from one feeding ground to another, the young would be protected from predators by walking centrally.

World of the sauropods

- Sauropods lived mostly during Jurassic times and then died out. Scientists think these herbivores may have evolved because of the lush vegetation that there was then. They certainly seem to have had the most enormous appetites.

This is evidenced by both large and small footprints found in the fossilized trackways that they left behind. Also left behind were huge piles of coprolites, the result of ingesting so much food.

It is even possible that they wandered into water at times to feed on aquatic plants, although due to their height it would have been a long way down to reach these.

LIVED: mostly in Jurassic times
SIZE: most were huge
WEIGHT: many tons
DISCOVERED: in North America, China etc

OTHER DATA: long necks and tail; herbivores; huge appetites; many swallowed stones, known as gastroliths, to help digestion; lived in herds; status: extinct

Their long necks, however, provided them with a very great advantage when it came to feeding from the treetops. There was no competition from dinosaurs other than the sauropods at such high levels.

Saurópods have been found in many parts of the world. One, which scientists have called Vulcanodon, was unearthed headless in Zimbabwe, Africa. We can only guess that greedy carnivores may have decapitated it during an attack and that the skull of their unfortunate victim became weathered and then disintegrated.

FACTFILE

The sauropod with the longest neck of all, extending to 49ft (15m), was MamenchIsaurus. It was found in what is now China.

Dinosaur extinction

No one is exactly sure how and why dinosaurs became extinct 65 million years ago. But scientists have come up with a whole range of interesting theories to explain their total disappearance from our planet.

In all, the many species of dinosaurs existed for about 150 million years. Then, suddenly, the last to have evolved all disappeared at the end of Cretaceous times. One possibility is that a deadly virus may have wiped them all out. Alternatively, it could be that they were unable to adapt to the changing climate, or that once more mammals had started to evolve, they ate large quantities of dinosaur eggs.

However, currently the most widely accepted theory is that a massive asteroid, as large as 6 miles (10km) in diameter, hit this planet with devastating effect. Scientists even think they know the place of impact – a site in Mexico, Central America, called Chiczukub, where a huge crater exists.

At this particular site they have found rocks full of substances usually associated with asteroids. Upon impact, lots of poisonous elements, such as sulphur, would have been released.

Once these were released into the atmosphere, sunlight was blocked out.

Deprived of light in this way, most of the plant life on Earth would have suffered, and herbivores would have starved to death as a result. Most of the carnivorous dinosaurs, in turn, would have had no prey on which to feed, so that they, too, became extinct in time.

ENDURING MYSTERY

However this does not explain why some life forms survived this catastrophe; nor why dinosaurs living in very distant parts of the globe died out, too. Scientists are therefore left with the task of finding a satisfactory answer to what remains an intriguing riddle. It may even be that they grew too large to mate!

FACTFILE

Some scientists believe that if, one day, they find dinosaur DNA, they may be able to recreate these extraordinary creatures.

DATA: dinosaurs died out about 65 million years ago, at the end of Cretaceous times. There seems to have been a mass extinction then, although not every life form disappeared. One prevalent theory is that an asteroid hit Earth, causing devastation as thick clouds of dust that were thrown up blotted out sunlight and prevented vegetation from flourishing; status: extinct

Pterosaurs

If humans had existed at the time of the dinosaurs and looked up into the sky, the chances are they would have seen flying reptiles swooping gracefully high above.

A typical pterosaur, the Ornithodesmus in this illustration seems to have been a skilled aviator. But what do scientists know about their means of flight? How did they launch themselves from the ground? And how did they keep in the air?

> **FACTFILE**
>
> Although they flew, pterosaurs are not related to birds. Instead, birds are descended from the dinosaurs.

Some palaeontologists think that the smaller pterosaurs would have launched themselves from trees, then relying upon air currents to soar. Others, however, believe they probably ran at speed for a short way before getting "lift." Either way, from models that have been made to assess how they flew, we know that they were probably able to flap their wings using powerful chest muscles. An average speed would have been around 24ft (7m) per second – a great deal faster than even an Olympic athlete can run!

But pterosaurs did not spend all their day flying, sometimes descending to rest and feed.

According to one school of thought, some pterosaurs may even have been able to float so that they could rest on water. Others, however, think that their wings were far too bulky for this. There is a theory, too, that some of the smaller pterosaur species may have been able to use their foot claws to hang upside down from branches to rest, much as bats do. It is even possible that a number may have been able to wrap themselves in their wings when they were suspended in this way.

No one has yet found a complete pterosaur egg, although remains of shell have been discovered.

LIVED: at the time of the dinosaurs
SIZE: from size of crows to size of light aircraft
WEIGHT: varied according to size
DISCOVERED: in many parts of the world

OTHER DATA: laid eggs; some lived in colonies; not related to birds; some had long tails, others had none; many had sharp claws; *say:* TE-ROH-SAWS; status: extinct

FIRST FIND

The very first pterosaur remains to be discovered were unearthed in Germany between 1767 and 1784. They were examined by an 18th century Italian naturalist, Cosimo Collini, who was mystified by them. They obviously did not belong to any known bird and yet the creatures seemed to have been able to fly. When the French anatomist Georges Cuvier, keeper of the Mannheim Natural History Collection in Germany, came to study an engraving of them, however, in 1801, he readily identified them as reptilian and named the specimen *Pterodactyle*, meaning "flight-finger." (Later, this particular type of pterosaur came to be known as *Pterodactylus antiquus* or "old flight-finger."

Scientists are, however, convinced for the most part that pterosaurs hatched from eggs and that some species may have formed nesting colonies. The nests themselves were probably hollows in the ground or constructed from foliage.

Pterosaur remains have now been found in many parts of the world, including East Africa, South America, England, the United States, the West Indies, France, Portugal, India, and China.

World of pterosaurs

- It could be that accidental discovery of pterosaur fossils, long before anyone knew what they were, led to legends about dragons. In fact, the Chinese use exactly the same word for a dragon as they use for a dinosaur – *kung-lung*.

Quetzalcoatlus

The largest pterosaur found to date, Quetzalcoatlus was the size of one of today's light aircraft and was named after Quetzalcoatl, the feathered serpent-god of the Aztec people of Central America.

Even the largest dinosaurs of Late Cretaceous times must have found the sight of this huge winged creature absolutely terrifying as it soared through the prehistoric skies.

Quetzalcoatlus had an amazing wingspan of approximately 39ft (12m.) But because of its hollow bones, in spite of its size, it probably only weighed as much as today's average adult man.

FACTFILE

Quetzalcoatlus was found in Big Bend National Park, Texas, USA, in 1971 by a student, named Douglas Lawson.

Most palaeontologists think it is likely that Queztalcoatlus was principally a marine-feeder and that it swooped down from the skies to suck up large quantities of fish, crabs, and molluscs. Its jaws were toothless, as far as we can tell from remains, which means that it would not have been able to chew its meals of seafood. Instead it could have swallowed everything that it ate in one great gulp.

There are some experts, however, who maintain that Quetzalcoatlus may also have fed on dinosaur remains at times, scavenging on any carcasses that it could find.

LIVED: in Cretaceous times
SIZE: wingspan of 39ft (12m)
WEIGHT: unknown
DISCOVERED: in North America

OTHER DATA: named after a South American god; probably fed mainly on fish; toothless jaws so swallowed its intake whole; *say*: KWETZ-AL-COAT-LOOS; status: extinct

This pterosaur may even have lived like a vulture, feeding mostly on carrion and pecking away at dead dinosaurs with its elongated jaws. But if it was a scavenger, its long, inflexible neck may have been something of a hindrance.

No one knows what colour this or any other pterosaur may have been. But it is likely any head crests may have been either a bright shade, or perhaps patterned in the mature creature. Quetzalcoatlus seems to have been among the very last of the pterosaurs, becoming extinct along with the dinosaurs at the end of Cretaceous times. Maybe it would have become even bigger if it had not died out then!

Rhamphorhynchus

The remains of a half-eaten early fish, discovered with the fossilized bones of the pterosaur Rhamphorrhynchus in Germany, provide a sure indication of the sort of diet it had.

Weighing up to 12 pounds (5.5kg) and with a wingspan of perhaps as much as 69ins (175cm), Rhamphorhynchus had sharp teeth along along its upper and lower jaws which were ideal for clamping down on any large fish it could find in the Jurassic waters, as depicted here.

LIVED: in Jurassic times
SIZE: wingspan up to 69ins (175cm)
WEIGHT: up to 12 pounds (5.5 kilos)
DISCOVERED: in Germany

OTHER DATA: pterosaur; pouch in jaw for storing fish; long tail; interlocking, sharp teeth; wings held by elongated bones of fourth finger; *say*: RAM-FOR-IN-KUS; status: extinct

There seem to have been several different species of this pterosaur – some as big as a swan, others more the size of seagulls. But palaeontologists believe that all may have had a special pouch of skin in their bottom jaw. Here they could have stored some of the fish that they caught as a first stage of digestion or even, once regurgitated into a softer, mushy consistency, to feed to their offspring back at the nest.

World of Rhamphorhynchus

- Rhamphorhychus had a very long tail with a diamond-shaped tip to it, made of soft tissue. Long tails are characteristic of some types of pterosaurs. Others, like Pterodactylus, *below*, had only what looked like a small stub for a tail.

BEAUTIFULLY PRESERVED

The remains of Rhamphorhynchus, which were first found in the limestone of the Solnhofen region of Germany, are in very good condition and show in great detail what this pterosaur was like. From them we know, for instance, that it had a sizeable head and a big sternum, or breastbone, that provided a large surface area for flight muscles. The elongated bones of a fourth finger on each forelimb, meanwhile, held its wings, which were strengthened by lots of tough fibres.

When it closed its jaws, the teeth locked together to form a sort of trap or cage, so that anything that Rhamphorhychus held in its mouth was unlikely to escape. The top jaw seems to have held as many as 20 long, slim teeth, while there were fourteen in the bottom jaw.

Pliosaurs

With huge jaws designed both for scooping up small fish and for slashing at larger prey, pliosaurs such as Kronosaurus, shown here, numbered among the most feared of sea creatures in prehistoric times.

Fossils of Kronosaurus unearthed in Queensland, Australia, show that this colossal marine creature was as long as 56ft (17m), and that its head alone extended to 8ft (2.4m), which is about the height of your ceiling.

From what palaeontologists have discovered so far, it was probably the largest of the pliosaurs and lived from Early Jurassic times right through to the end of the Cretaceous period when, along with the dinosaurs, it disappeared.

Just look at all those nasty teeth! These helped to make short shrift of slashing at victims, which Kronosaurus would then hurriedly devour. Such huge creatures had gigantic appetites.

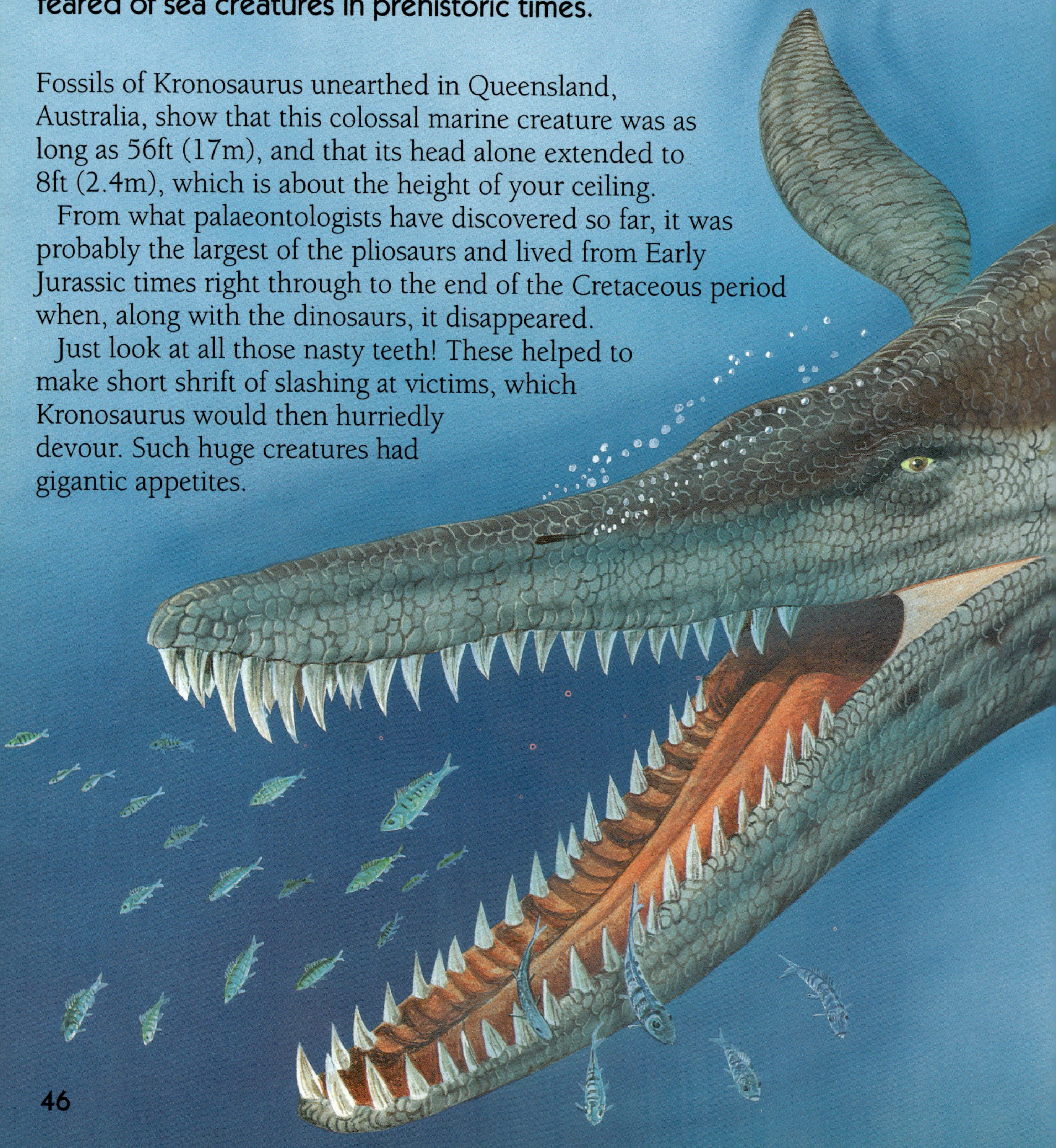

LIVED: from Early Jurassic to Late Cretaceous times
SIZE: up to 56ft (17m)
WEIGHT: many tons
DISCOVERED: in Australia and Europe

OTHER DATA: thick necks; huge heads; sharp teeth in massive jaws; mighty predator; would have attacked marine creatures of all sizes; *say*: PLEE-OH-SAWS; status: extinct

World of the pliosaurs

- Kronosaurus is named after a Greek god, Kronos. Like all pliosaurs, it was not a dinosaur – dinosaurs did not swim, remember – but a marine reptile and survived as a species for about many millions of years.

Other pliosaurs included Pliosaurus, which measured about 22ft (6.7m) in length, and Peloneustes which was only a fraction of this size.

All had closely packed, sharp teeth that were designed for tearing huge chunks of flesh out of their prey. Larger specimens may even have gone after creatures such as the gigantic long-necked plesiosaur that you will meet when you turn to page 50.

Placodonts

With a name meaning "flat teeth," this group of creatures was among the first reptiles on our planet to move into the water more than 215 million years ago. Several of their fossils have been found in Italy.

Not entirely aquatic and in some ways more suited to getting about on land than in the water, placodonts nevertheless probably spent a good part of their day in the warm shallows during Triassic times.

Here, as remains found in their stomach region and also their mouth structure show, they feasted on shellfish of many kinds which they would tear from the seabed. Their powerfully muscled, shovel-shaped top and bottom jaws and six sharp, peglike front teeth (incisors) would have come in useful for feeding on such fare.

Then their 14 blunt, flat, crushing back teeth, which were covered with thick enamel, would have ground up the seafare, producing a mush that was easy to digest.

One placodont, known as Placodus, had a longish tail and stout body, but a short neck. Scientists think it could even have possessed a functioning third eye which would have provided a degree of protection against predators. This either enhanced vision or may have assisted with orientation.

Other placodonts, however, such as Placochelys and Henodos (dug up from Triassic strata in Germany), had particularly strong body armour as their main means of defence.

Placodus had long tail which helped to propel its high-backed body along underwater, as you can see in the depiction shown *opposite*.

Mysteriously, there is no evidence of any of the placodonts surviving beyond the Late Triassic period. So they probably disappeared about 200 million years ago, along with other creatures that vanished in the widespread mass extinction that occurred then.

One theory suggests that this extinction of so many different forms of animal life may have been due to the rise of the dinosaurs; another, that a marked shift from heavy rainfall to a much drier type of climate could have been to blame.

World of the placodonts

- Whenever placodonts ventured into the water, they would have been in search of food.
- Molluscs, brachiopods, and crustaceans would have been the staple part of a placodont's diet.

LIVED: in Triassic times

SIZE: Placodus was 6.5ft (2m) long

WEIGHT: unknown

DISCOVERED: in Germany and Italy

OTHER DATA: long tail; powerful jaws; peglike incisor teeth; ate mostly shellfish; short neck; stout body; Placodus may possibly have had a third eye; status: extinct

Plesiosaurs

It was 15-year-old Mary Anning who first discovered Plesiosaur remains at Lyme Regis in southern England, near to her home, in 1814. It is still a favourite fossil-hunting area.

Mary Anning's father sold fossils for a living, and it was no doubt this background that gave her an early interest in palaeontology. She knew the value of fossils, as we can tell from the fact that she sold the plesiosaur remains she found in the chalky cliffs for what would be over £100,000 ($150,000) in today's terms. But money was not Mary Anning's main concern. She was passionately interested in collecting prehistoric remains and in piecing them together, even though she had no academic training.

Plesiosaurus lived in the water and the only time it would have come ashore would have been to lay its eggs. Otherwise, it remained in the water, feeding on passing fish. It may even have swum into shoals intentionally, attempting to scatter and grab some.

World of the plesiosaurs

- Plesiosaurs are sometimes described as ribbon reptiles because of their long ribbonlike necks. Astonishingly, many had necks that made up over half their total body length.

LIVED: in Jurassic and Cretaceous times
SIZE: up to 45ft (14m)
WEIGHT: unknown
DISCOVERED: in England and elsewhere

OTHER DATA: very long necks; laid eggs on land; barrel-shaped bodies; interlocking teeth; could not breathe under water so came up for air; *say:* PLEEZ-EE-OH-SAWS; status: extinct

The longest plesiosaur found so far was one discovered in the state of Kansas, USA, as well as in Japan. Elasmosaurus had such a lengthy neck that it was even described as a "snake threaded through the body of a turtle." In all, it had seventy-one neck bones – an enormous number in comparison the seven that we have!

When it swam, a plesiosaur such as Elasmosaurus would have held its neck completely rigid and straight in front of it, like an elongated arrow. But when a plesiosaur came to rest, the neck could be coiled into a loop, or maybe even two loops, as some palaeontologists have suggested.

Front flippers would have been used to propel a plesiosaur through the water, while the back flippers were for steering or helping the creature come to a standstill.

Plesiosaurs had no body weapons and must often have been targets for vicious, bulky marine predators. Nevertheless, they survived to the end of the Cretaceous era, by which time some were well over 40ft (12m) in length.

FACTFILE

Plesiosaurs had small lungs and could not breathe underwater, so they had to come up for air every now and then.

Early mammals

Alongside the dinosaurs lived several types of small mammals, as shown across these two pages. Larger mammalians had yet to evolve at this point in prehistory.

The odd creature shown *left* is a Tritylodont which dates from Triassic times. It was about 3.5ft (just over 1m) long and had unusual teeth, as you can see. In fact, its name means "three-knobbed tooth." Behind its larger front teeth, those teeth in its lower jaw had two knobs, and those teeth in the upper jaw had three.

Scientists have only been able to guess at its appearance from its fossilized skull and teeth, as few remains have been found.

Triconodon dates just from later Jurassic times and was only the size of one of today's average pet cats. Still, it seems to have been predatory, and would have hunted all sorts of small creatures on which to feed. Some palaeontologists think that it may even have been able to climb trees to seek out its prey or to avoid an enemy, and that it may have feasted on dinosaur eggs.

LIVED: some had evolved by Triassic times
SIZE: some early mammals were the size of rats
WEIGHT: not heavy
DISCOVERED: in North America and elsewhere

OTHER DATA: as mammals, did not lay eggs but gave birth to live young; larger mammals evolved later; marsupials existed among the early mammals; status: extinct

Remains of these long-jawed creatures have been dug up in what is now Europe, where they lived in constant fear of being trampled on by far larger beasts.

Looking very much like an opossum, Alphadon, *right*, lived in what is now North America. It was a marsupial and so had a special sort of pouch for carrying its young.

FACTFILE

Early mammals probably had a hairy body covering, and their whiskers were highly sensitive sensory organs.

Megazostrodon, *below*, dates from the Triassic period, as does Tritylodont, *top right*. It was tiny, shrewlike, and only grew to about 5ins (13cm) long. Nevertheless, it was a probably principally an insectivore, only going out at night when there was less risk of predators. Its remains were first found in South Africa.

But how can scientists tell that a creature was a warm-blooded mammal from such incomplete fossils? In many instances, there is a structure showing the nose was divided from the mouth. This allowed feeding and breathing at the same time, something which most modern reptiles simply cannot manage to do.

Draco

Although the dinosaurs had their feet planted firmly on the ground – they neither swam in the seas nor flew in the air – some of their smaller reptilian cousins developed wing-like flaps to help them glide from place to place. In the illustration *below*, you can see one that is leaping away to the safety of another branch as a hungry sauropod feeds on the leaves of the very same tree. But the Draco need not have worried – sauropods, as you know, did not eat flesh.

A whole host of prehistoric winged lizards could make short excursions through the air, but they could not really fly. The greatest distance that a creature known as Draco could cover, for instance, was about 200ft (60m), but this was unusual.

TAKING OFF

Most winged lizards could leap about 40ft (12m), a fair distance for a creature such as Early Triassic Coelosauravus, which was only 16-18ins (40-45cm) long, to propel itself between trees, or from a lakeside take-off point up to a higher ground level.

Fossilized remains of winged lizards, such as Late Triassic Icarosaurus – named after the Greek god who made wax wings in order to be able to fly – show that they had four limbs and, a very scaly skin.

LIVED: at the time of the dinosaurs
SIZE: wing membranes up to 13ins (33cm) across
WEIGHT: not at all heavy
DISCOVERED: in several parts of the world

OTHER DATA: these winged lizards caught insects; leapt and then glided through the air; lengthy tails; scaly skin; could escape danger by gliding away at speed; status: extinct

They also had a lengthy tail and elongated, protruding ribs covered by a thin membrane which supported them in flight. Winged lizards are able to spread out these flaps by expanding their rib cages.

World of Draco

- Remains of *Draco volans*, meaning "flying dragon," have been found in Malaysia.

Late Triassic Kuehnosaurus, meanwhile, had swept-back flaps, like a modern fighter plane, which it could fold neatly against its flanks after landing, as could Icarosaurus and Daedalosaurus. This manoeuvre prevented their flaps from being caught as the creatures slithered through dense foliage.

The winged lizards used their unusual aerial skills to catch insects – which were either surprised and gobbled up when the lizard suddenly landed, or snatched from the air during flight. They would also glide in the attempt to get themselves out of trouble if a predator approached. But many of the smaller, meat-eating dinosaurs would have made short work of a winged lizard.

However, for one species at least, the ability to glide and also change colour to camouflage itself allowed it to survive until the present day. Indeed, these relatives of Draco still live in the jungles of South-East Asia, where they are known as Flying Dragons.

Sarcosuchus

Throughout Jurassic and Cretaceous times, there were pockets of land that were constantly submerged in water. These swamps provided a lush habitat for hundreds of semi-aquatic creatures. Among them, there were early species of crocodiles, such as Sarcosuchus.

Warm sunlight dappled the murky waters of the Early Cretaceous swamp, making it even harder for the wandering dinosaur, an Ouranosaurus, to watch for lurking enemies as it selected a tasty horsetail on which to munch. Sarcosuchus, lurking in the reeds, had no such problem.

POWERFUL PREDATOR

Few creatures passing this massive, 40ft (12m)-long crocodile-like creature were likely to escape its snapping jaws. Even majestic, sail-backed Ouranosaurus would frequently have been at risk.

Early turtles, too, in spite of their hard shells, would have been sure to keep their distance, for fear of being flipped over by Sarcosuchus. If this happened, their soft underbellies would be exposed to the crocodile's scything teeth, and survival was unlikely.

Swamplands abounded in prehistoric times and usually arose from waterlogged river deltas, such as those fanning out over certain parts of North Africa.

LIVED: in Cretaceous times
SIZE: up to 10ft (3m) long
WEIGHT: as today's crocodiles of similar size
DISCOVERED: in Niger, Africa, and elsewhere

OTHER DATA: long, snapping jaws; frequented swamps; would attack early turtles and some dinosaurs; *say*: SARK-OH-SOOK-OOS; status: extinct

They were gloomy places, where forests of high cypresses, ginkgoes, and seed-ferns blotted out the light. But along the water's edge, reed-like horsetails, mosses and lichens flourished, as did certain flowering plants that the world saw for the first time in the Cretaceous period.

OTHER CROCODILIANS

These ancestors of today's crocodilians were undoubtedly kings of the prehistoric swamplands. Among them, in addition to Sarcosuchus, were such mighty creatures as heavily armoured Rutiodon, which preyed on fish and other reptiles.

Later crocodile-like reptiles, such as Goniopholis, evolved into creatures closely resembling the large crocodiles infesting modern swamplands.

At up to 10ft (3m) in length with long, snapping jaws, scaly backs, and clawed feet, these crocodilians were ferocious predators, just like the crocodiles of today, and could be relied upon to fight to the death.

FACTFILE

Like dinosaurs, crocodilians such as Sarcosuchus are classified as archosaurs, meaning "ruling reptiles."

Ammonites

It is easy to see how ammonites came to get their name. Because they had coiled shells, they were named after an Egyptian god, Amun, who was always depicted with curling ram's horns.

First appearing during the Jurassic era but becoming extinct by the end of Cretaceous times when the dinosaurs also disappeared from Planet Earth, ammonites were marine creatures with the most fascinating coiled shells. The shell itself was always divided into distinct chambers separated by walls which scientists know as *septa.* The animal itself only lived in the last of these. The other chambers, meanwhile, were filled with a sort of gas or fluid. The ammonite could control its movement in the sea by careful regulation of these.

Ammonites of many different kinds have been found almost all over the world, including Europe, the United States, the Himalayas, as far north as Greenland, Africa, Japan, Indonesia, South America, Russia, and New Zealand.

World of the ammonites

- Ammonites were once thought to provide a cure for infertility and baldness if worn as lucky charms; but, of course, we now know this does not work.
- Ammonites were once known as snakestones because legend had it that their coiled form showed they were serpents that had been turned into stone.

LIVED: in Jurassic and Cretaceous times
SIZE: from tiny to 3ft (0.9m) in diameter
WEIGHT: varied according to size
DISCOVERED: almost worldwide

OTHER DATA: live creatures inside the shells; movement controlled by contents of chambers inside the shells; fed on plankton and marine vegetation; status: extinct

WHORLS WITHIN WHORLS

The shells of ammonites feature a few whorls or circular coils, and it is these that give the ammonites their distinctive shape. Sometimes the shells would also feature spines and other small projections which would give the creature inside added protection from a whole assortment of predators which might otherwise have found it easy to crack an ammonite open. Some ammonites have even been found with puncture marks.

The ammonites themselves probably fed on minute plankton and marine plants.

FACTFILE

Some ammonites were very small, but others were the size of your arm in diameter, or even bigger.

They were fleshy creatures with large eyes and many tentacles. In times of danger, these tentacles could be withdrawn into the shell. If the ammonite wanted to move quickly and escape from a predator, it could squirt out a stream of water which would send it shooting backwards to a safer point at a terrific speed.

Coelacanths

Sometimes we are tempted to believe that every prehistoric species is dead and gone. But, amazingly, there are still creatures alive today whose ancestors existed at the same time as the dinosaurs.

It had been a long day, but the fishermen were satisfied. The waters of the Indian Ocean that surrounded their native Comoros Islands, off the coast of south-east Africa, teemed with fish; and, as usual, they would be putting into port at East London, South Africa, with their small boat groaning under the weight of their catch.

On that day in December 1938, however, a peculiar specimen caught their eye. The 6ft (1.8m)-long creature thrashing for dear life out on the quayside was unlike any fish they had see before. Perhaps the young lady scientist from the museum would be interested, they wondered.

As it turned out, Marjorie Courtenay-Latimer could hardly contain her excitement when they presented the fish. Lying in front of her was a creature that she had only seen before as a fossil, frozen in rock that was millions of years old. The shape of its body, fins, and tail all matched those of a prehistoric coelacanth, assumed to have become extinct since no remains ever found were less than 60 million years old.

FURTHER FINDS

When another was found fourteen years later, in 1952, it was examined by the Capetown fish expert Professor J. Smith, who announced that the coelacanth was indeed a living fossil, surviving unchanged for 400 million years.

How, then, had coelacanths survived, when their prehistoric companions had long since perished? This riddle puzzled scientists until 1987, when the German marine biologist Hans Fricke used a submersible craft down to explore the ocean bed around the Comoros Islands. There, in very deep water, he discovered a living coelacanth calmly exploring a crevice in some rocks. Oddly enough, there did not seem to be any marine life round it, although Fricke knew from remains of stomach contents found in coelacanth fossils that its ancestors dined on squid and other fish.

FACTFILE

The coelacanth has been nicknamed "old fourlegs" because of its limb-like pectoral and pelvic fins.

LIVED: still survives, as a "living fossil"
SIZE: 6ft (1.8m) long
WEIGHT: heavy as bulky for its length
DISCOVERED: in the Indian Ocean

OTHER DATA: first rediscovered in 1938; appearance just as its fossils show it was 400 million years ago; hid from other marine life; *say:* SEEL-AH-CANTHS; status: rarely seen

The key to the coelacanth's survival, he claimed, was the reclusive life it led. Being so bulky and probably slow off the mark, he reasoned, coelacanths could not compete for food with more streamlined, and therefore faster, fish. So they retreated to sparsely populated depths, where the faster fish did not bother to go.

Phororhacos

About 15-20 million years ago, in what is now South America, there lived a huge flightless bird. It was more than 9ft (2.7m) in height and had powerful legs that helped it to run at great speed. Its large, hooked beak, meanwhile, was ideal for tearing at raw flesh.

LIVED: 20 million years ago, in Miocene times
SIZE: 9ft (2.7m) in height
DISCOVERED: in Patagonia, South America
DIET: mainly a meat-eater

OTHER DATA: a bird but unable to fly; rudimentary wings only; fast runner; large skull, up to 2ft (60cm) long; probably laid very large eggs; *say*: FOR-OR-HARK-OOS; status: extinct

Scientists know that Phororhacos almost certainly ate meat because its large claws and beak point to the lifestyle of a typical carnivorous bird. The illustration, shown *opposite*, provides an indication of what it would have looked like as it caught a small creature. It would definitely have been able to devour this lizard in one gulp.

Phororhacos had a very large skull for a bird, measuring up to 2ft (60cms) long; but in other respects it was very much like the ostriches we know today, which, of course, also cannot fly.

Most of its time would have been spent running over the plains of Patagonia, looking for its next meal. But how did Phororhacos itself manage to escape the clutches of fiercer beasts of prey? Quite simply, at that time in that part of the world, scientists think, there were no huge carnivores, and so creatures such as Phororharcos were not at risk of ending up as some other predator's next big meal.

An even earlier and equally curious flightless bird was Diatryma, which lived in what is now the North American continent, in the area of the state of Wyoming, as long as 50 million years ago.

An almost complete skeleton of Diatryma can be seen in the Natural History Museum of New York. They, too, were tall and had strong legs to facilitate running. The beak of this bird, however, was more parrotlike than the beak of Phororhacos, and its skull was not quite as long. It, too, had only rudimentary wings so that we can be sure these were not strong enough for flight and that this bird lived on land.

It is unlikely that either bird lived in large groups, and certainly there has been no discovery of several complete fossilized skeletons in one spot. It could have been that they lived singly, only coming together for mating.

Both these flightless birds no doubt laid the most enormous eggs, just as ostriches do. They would have been many times the size of chicken eggs.

World of Phororhacos

- The period in which Phororhacos lived in what is now South America is known as the Miocene.
- The head of Phororhacos was even larger than that of today's horses – a most unusual feature for any bird.
- Phororhacos may have included vegetation in its diet, as well as meat.

Glyptodon

Protected by a tough shell-casing, Glyptodon could certainly stand up for itself against any predator that threatened, and was undoubtedly one of the strangest-looking creatures the world has ever known.

Picture the scene as, with a mighty roar, the angry sabre-toothed tiger leapt up to confront the creature that had disturbed its snooze. But as soon as it saw the trespasser, the sabre-toothed tiger thought again. Both the very size of the beast and the extent of its body armour were awesome.

This strange animal was a Glyptodon, a forerunner of today's armadillo. A bony shield covered its back, while its head was protected by a structure resembling a helmet.

There were spikes on its tail, too. So, all in all, its soft underbelly was very well-protected from attack. But although it looked fierce, Glyptodon was not a carnivore, feeding instead on the grasslands of prehistoric South America rather than on other animals.

COOLER TIMES

But as the world's climate got cooler, the nature of the local vegetation started to change, and Glyptodon suffered. It now began to meet with serious competition for food, and the population soon diminished. Glyptodon finally died out 50,000 years ago, well before the arrival of modern humans.

FACTFILE

Glyptodon shells grew to be very large, and were exceedingly thick and tough, extending to a skull-covering, too.

LIVED: until Pleistocene times
SIZE: 11ft (3.3m) long
WEIGHT: 4,400 lbs (2,000 kilos)
DISCOVERED: in South America

OTHER DATA: huge shell covering; spikes on tail gave added protection; soft underbelly; helmet-like protection for its head, too; herbivore; *say*: GLIP-TOH-DON; status: extinct

It was to have a strange memorial, for the remains of its huge and hard-wearing shell were actually turned into shelters by humans.

For early South American peoples, these shells must have made small but extremely sturdy dwellings.

Smilodon

A type of sabre-toothed tiger, smilodon was a predatory carnivore with huge canine teeth that were as sharp as daggers.

Sabre-toothed cats must have been among the most savage killers of all time. With the help of their very powerful neck muscles, they would viciously dig their awesome fangs deep into the throat or belly of their unfortunate quarry. It would have been a most horrific sight.

Some scientists think that it was the blood and guts of the victims that the sabre-tooths were also after, in addition to the animals' flesh. These fierce creatures may even have found it necessary to kill several time every single day in order to get sufficient food on which to survive.

FACTFILE

Packs of smilodons probably migrated over great distances, using land bridges to cross water, as did bison and mammoths.

SURPRISE DISCOVERY

Los Angeles, in the state of California, USA, must today be one of the busiest cities in the world. It is constantly abuzz with business activity, and its freeways are thick with traffic. Meanwhile, its suburb, Hollywood, is the home of the movie industry. A Disney theme park is also not far away in the district of Annaheim.

So who would have thought that in a nearby place called Rancho La Brea, palaeontologists would make a major discovery – the bones of as many as two thousand sabre-toothed tigers!

The dig started in 1913 and lasted about 20 years. In that time, it yielded the fossilized skeletons of many types of animals and birds – everything from antelopes, ground sloths, elephants, and bison to hawks, deer and wolves. There were mammoths, too, and even camel remains.

LIVED: prior to 11,000 years ago
SIZE: tiger-sized
WEIGHT: as a modern tiger
DISCOVERED: in North and South America

OTHER DATA: mammal; huge fangs; went for throat or belly of its victims; probably responsible for disappearance of many herbivores; *say:* SMILE-OH-DON; status: extinct

It seems that thousands of years ago, in Pleistocene times, these creatures had all become stuck in a large muddy pit. Oil had been forced up from the lower rock strata in that area, forming very sticky tar, from which there was no escape.

What probably happened was that herbivores came to drink at what they thought was a lake but did not stay at the edge. Instead, they entered the water-covered pit, did not see the mud beneath the surface, and got stuck. Carnivores, like smilodon, then came in after them and also sank into the mire. This must have happened time and time again, over centuries perhaps, to judge by all the skeletons that have been found in this area.

World of smilodon

- The world's last sabre-tooths perished around 11,000 years ago, probably due to temperature and sea level changes.

Cave bears

When a cave bear left its hide-out and ventured into the outside world, many other creatures must have fled for their lives. Yet it was rarely predatory and might well have left them alone, particularly in summer when it usually fed on vegetation, as evidenced by its type of teeth.

About the same size as one of today's grizzly bears and perhaps equally as fearsome, cave bears were nevertheless driven out of their homes by our ancestors throughout the last Ice Age. Indeed, competition over caves as places of shelter between these bears and various hominids, particularly the Neanderthals and the Cro-Magnons, is said to have been extremely fierce. Cave bears lived in packs, and so it must have been very difficult to evict them. In just one Austrian cave, scientists have even found as many as 30,000 cave bears. Of course, they did not all live there at one time; but this discovery shows that the bears used that cave over very many generations.

World of the cave bear

- Cave bears were hunted by early humans but died out mainly because, largely herbivores, they could not find sufficient food to sustain themselves during the last Ice Age. Finally, the last of them probably starved to death.

FOSSIL EVIDENCE

And how do scientists know that they fought with hominids? Evidence is there in the form of spear marks on their fossilized bones, and also the broken tips of these weapons.

During Pleistocene times, cave bears were to be found in many parts of what is now Europe – Poland, France, Switzerland, Germany, Austria, England, and the former Yugoslavia, for example; and they have been unearthed in Russia and China, too. Indeed, it is estimated that the last cave bears probably died in the region of Yugoslavia about 10,000 years ago. It may even be that the bones of hominids found in some caves belonged to individuals who had been killed by such highly territorial cave bears.

These bears were born and hibernated in caves, and often died there, too. Fossilized footprints also show that bears once lived in certain caves. The walls, meanwhile, have marks that were made as they scratched at the surface in the attempt to keep their claws sharp. The world may not see the like again of these creatures, finally driven to extinction by loss of feeding grounds.

LIVED: in Pleistocene times
SIZE: as big as today's grizzly bears
WEIGHT: as heavy as today's grizzly bears
DISCOVERED: in many parts of Europe and China

OTHER DATA: fought with early humans; evidence found in caves; very territorial; fossilized footprints discovered, too; disappeared during last Ice Age; status: extinct

Giant ground sloth

No one knows what colour this mighty, lumbering creature was. Remains indicate, however, that it was covered in luxuriant fur.

As shown in this illustration, giant ground sloths could easily reach up and pull down a tasty-looking branch. This was the one place that few other animals could reach, so it meant that the sloths had plenty of food.

Related to today's far smaller, tree-dwelling sloths, some were so huge that, when they reared up on their back legs, they stood all of 17ft (5.2m) tall – the same height as today's giraffes.

These strange-looking creatures became extinct around the time of early humans. They may have been hunted by us; but it seems that our arrival was not the main cause of their extinction.

Instead, it is likely to have been the changing climate and landscape.

A few years ago, fur and dung belonging to the giant ground sloth were found in a cave in South America. Some people thought it meant this creature was not extinct after all, but tests showed the remains to be remarkably well-preserved and yet many thousands of years old.

World of the giant ground sloth

- This prehistoric creature's remains have been dug up in North and South America, where it probably lived alongside such creatures as glyptodons during Tertiary times. It could have weighed as much as several tons.

LIVED: until 11,000 years ago
SIZE: 17ft (5.2m)
WEIGHT: not known, but possibly several tons
DISCOVERED: North and South America

OTHER DATA: mammal; its fossilized skeletal remains, as well as hair and dung, discovered; herbivore; may have lived part of its time in caves; status: extinct

Woolly rhino

The thick, reddish-brown coat of the woolly rhino would no doubt have helped to protect this tough creature from the very severe weather conditions of the regions it inhabited prior to extinction.

But when up against an enemy, it no doubt charged at a terrific pace.

This rhino stood nearly 5ft (1.5m) tall at the shoulder and was approximately 10ft (3.5m) in length. Many remains have been found in Siberia and Poland.

Easily recognisable because of its thick, shaggy body covering and the handsome pair of horns – the front one a lot longer than the other – that it sported on its snout, the woolly rhino lived alongside the mammoths that existed in Late Pleistocene times.

There are unlikely to have been fatal encounters between them, however. The woolly rhino may have looked ferocious, but it would not have attacked for food since it was strictly an herbivore.

In any event, this bulky animal probably did not move at great speed, lumbering along just like the rhinos of today. The deep snow of the Siberian steppes would no doubt have slowed it down even more.

Whenever possible, it would feed greedily on huge quantities of grass.

This helped the woolly rhino of 400,000 – 25,000 years ago to build up reserves of body fat and energy to provide maximum chances of survival in less favourable times.

World of the woolly rhino

- Native to Siberia and then migrating to parts of Europe, the woolly rhino died out for the most part at the end of the last Ice Age. We know what it must have looked like from skeletal remains, as well as the cave paintings of our ancestors.

LIVED: prior to 10,000 years ago
SIZE: 10ft (3m) long
WEIGHT: probably that of a rhino of today
DISCOVERED: in the Ukraine, Siberia, Poland

OTHER DATA: large front horn, smaller rear horn; herbivore; could charge furiously to defend itself; shaggy body covering kept it warm in icy conditions; status: extinct

Known by the scientific name *Coelodonta antiquitatis* (*say:* SEEL-OH-DONT-AH ANT-IK-WIT-AHT-IS), the woolly rhino became extinct about 10,000 years ago when there were great changes in the mammal population of all of the continents.

Some scientists think that the climate changed, becoming warmer as the ice sheets retreated. Others, however, hold that as the human population increased, so more and more mammals were hunted for food.

In the Palaeontological Museum in Krakow, Poland, there is a magnificent specimen of a female woolly rhino, unearthed from muddy deposits in the Ukraine. Amazingly, even its soft tissues are in remarkably good condition because the whole carcass had become preserved, over 10,000 years, in deposits of oil and salt. Interestingly, its front horn is flattened from side to side, unlike the horn of today's rhinos. It was probably used to dig through snow to get at the grass beneath.

Dinictis

Just imagine the agony of the early horse in this picture as it was pounced upon by a savage member of the cat family way back in prehistoric times! A horrible death must soon have ensued.

In North America, about 30 million years ago, generations of early horses, and many other animals, too, must have fallen victim to this predator.

Dinictis knew instinctively how best to dispose of its quarry; and such was the ferocity of its attack that the unfortunate beast destined to be its next few meals would have died writhing in agony within just a few minutes.

LIVED: 30 million years ago
SIZE: 3.5ft (1m) long
WEIGHT: unknown
DISCOVERED: in North America

OTHER DATA: fierce carnivore; sometimes called a false sabre-tooth because not as large as these creatures and had smaller but still very nasty fangs; sharp claws; status: extinct

FACTFILE

Most of the animals that lived at the time of dinictis, in the third period of the Tertiary era, are in fact now extinct.

Although dinictis only stood about 2ft (0.6m) in height at the shoulder and had a body length of just 3.5ft (1m), it was still a formidable beast and would have attacked at every possible opportunity.

You may perhaps think it strange that an animal of only this size would dare to attack a horse. But prehistoric horses were only small and dinictis was wild in every respect. Few creatures of any kind were safe when it was out and about on the prowl.

When it went for its prey, dinictis would first dig into the victim with its nasty claws. Then it would use its fangs to cut out a chunk of flesh and leave the victim to bleed to death. Dinictis could now feast on the meat along with its mate and any of their offspring.

Smaller than smilodon and other, later sabre-toothed tigers (turn to pages 66-67 to read about them, too), dinictis also had powerful upper teeth, but these were substantially reduced in size. In fact, dinictis is sometimes called a *false* sabre-toothed cat because its fangs hardly measured up to those of the sabre-toothed tigers. However, they were still extremely well suited to grabbing and biting savagely at prey.

KEEN SENSES

Dinictis would probably have had very keen eyesight and hearing, and a good sense of smell, too. No one knows, however, what its colouring and skin markings would have been like, if indeed there was any patterning; and we can only guess that it would have had long, sensitive whiskers.

We can assume, too, that it would have made nasty squealing noises as it attacked, and that it did so with terrifying force.

World of dinictis

- Remains of dinictis have been found by palaeontologists in the North American states of South Dakota and Nebraska.
- Dinictis had a longer tail than most of the sabre-toothed cats, which are also known as *machairodontids*.

Mammoths

These elephant-like creatures became finally became extinct around 4,000 years ago, having lived on Earth for over 2000,00 years. Their remains have been found in Europe, North America, and Siberia.

Why mammoths became extinct is still largely a mystery, but scientists have several ideas about what may have happened to bring about their demise. Some have suggested that, towards the end of the last Ice Age, about 10,000 years ago, there may have been several dreadfully cold spells when the mammoths could not find food and so starved to death.

FACTFILE

Woolly mammoths weighed about as much as 80 large adult men today and were covered in shaggy, dark hair.

BOGGED DOWN

Another theory is that, as the weather gradually grew warmer, the ice began to melt. So, as the ground got softer, the heavy mammoths sank into bogs, and became trapped in the mud. Unable to escape, they died there. Their bodies may then have frozen in the ground as the weather changed yet again.

In some places, such as Rancho La Brea in California, USA, the remains of mammoths have been found in gravel and tar pits. So, almost certainly, they did blunder into them in this way. Hunters may even have driven them into such pits.

But other scientists, however, think that, in warmer weather, a number of the mammoths may often have become short of water. They may therefore have crowded along the shrinking water holes, lakes, and rivers, eating all the available food that was to be found nearby.

As they gathered together in increasing numbers, they may have attracted large carnivores which were in search of a meal. Weak from lack of food and water, the mammoths would not have been able to fight off these aggressive, hungry predators. Disease may also have spread rapidly through such large herds, causing many deaths.

DEEP-FROZEN

The very best mammoth discoveries have been made in Siberia, where the remains of over 4,500 of these creatures have been found. Here, the frozen ground has acted like a giant deep freezer, preserving the animals trapped in it. Scientists even think that about 500,000 tons of tusks may still be in the ground there. These are frequently stolen to be carved into ivory ornaments.

LIVED: prior to 4,000 years ago
SIZE: up to 11ft (3.4m) tall
WEIGHT: up to 6 tons
DISCOVERED: in Europe, North America, Siberia

OTHER DATA: herbivore; huge tusks; some found well preserved in Siberia under the permafrost; hunted by early humans for their meat and tusks; status: extinct

World of the mammoth

- As herbivores, mammoths spent most of their time looking for vegetation to eat, especially grass. Some mammoth remains are so well preserved that scientists can even examine the stomach contents.
- Large animals like the mammoths survived well in cold temperatures because their size ensured that there was a reduction in heat loss through the skin. A thick layer of fat also helped to keep them warm.

Alticamelus

Looking much like a cross between a camel and a giraffe, Alticamelus is known to have roamed North America, where remains have been unearthed in Colorado. In view of its height, it could browse safely on the treetops without competition from other herbivores.

Today there are two types of camels. One is the single-humped dromedary; the other, the two-humped Bactrian which, in the wild variety, is now severely endangered. (You can find out more about wild Bactrian camels if you turn to page 214). Prehistoric Alticamelus had features in common with today's camels, but there were some differences.

LIVED: in Miocene times
SIZE: 10ft (over 3m) tall
WEIGHT: unknown
DISCOVERED: in North America

OTHER DATA: name means "tall camel;" fossilized tracks found; no trace of a hump; long legs; long, slim neck; walked like modern camels; status: extinct

With a name meaning "tall camel," Alticamelus lived in the Miocene period, from about 10-5 million years ago. As you can see from the illustration *opposite*, it had very long legs and a lengthy, slim neck. Its tail was short and, unlike a true camel of today, it only had a slight bulge to its back, not a hump as such.

It did have hooves, however, and would have stood about 10ft (over 3m) in height. But its neck alone comprised half of this.

It may be hard to believe that camels were once part of the natural fauna of what is now North America; but most of their evolution did indeed take place in this part of the world.

Only later did they reach South America where they are represented by today's llamas. The natural habitat of today's true camels, meanwhile, is North Africa, Mongolia, China, and the Middle East, of course.

FACTFILE

Since it had no hump at all, Alticamelus probably could not store nutrients over long periods like a true camel.

FROM SMALL BEGINNINGS

Alticamelus probably started to evolve at a far smaller size, perhaps that of a sheep; but gradually, over millions of years, these creatures became far taller and developed highly elongated necks. They would have started to walk with a lumbering gait like modern camels, too.

Fossilized tracks that date from Miocene times show us that there were certainly creatures that first moved both right limbs together and then both left limbs as they walked or ran at a fair pace over what must have been vast areas of grassland. Only relatives of the giraffe and the camel walk like this.

No one is sure why some camels of Miocene times developed long, mostly toothless snouts, apart from a couple of incisors at the front; but it may have been to give them extra height so that they could reach up to the foliage at the top of even taller trees.

World of Alticamelus

- Alticamelus is sometimes given the alternative name Aepycamelus. Fossilized tracks found at a place called Copper Canyon at Death Valley Monument in the United States, measuring about 8ins (20cm) in length and 6ins (15cm) in width, may well have been made by herds of Alticamelus.

Mesohippus

Known to have inhabited what is now the North American state of Nebraska about 30 million years ago in Middle Oligocene times, Mesohippus was a miniature type of horse, living in herds. It was part of the evolutionary line of the horses we know today.

A great many remains of prehistoric horses have been unearthed by palaeontologists, and this has made it possible for them to assess how the horse may have changed over many millions of years.

The first horses are thought to have appeared in North America about 54 million years ago, after which they spread, in a series of migrations, to Eurasia and South America. Scientists think that they probably travelled to Eurasia via a land corridor that existed where a sea, the Bering Strait, is today.

Hyracotherium, also known as Eohippus and a hare-sized ancestor of the horse, lived in Eocene times in North America and Europe. It was probably a forest-dweller, living singly and browsing on low foliage.

Only in Oligocene times did horses the size of a large dog start to appear. Among these tiny horses was Mesohippus, a name with the meaning "intermediate horse." You can see a herd of them nibbling peacefully at grass in the illustration, shown *right*.

Each of the four limbs now had three toes. The central one was larger than the other two and so bore most of the body weight. Then, over time, horses evolved with hooves like those we are familiar with now.

But the evolution of horses did not proceed in a straight line. Many different types evolved and then died out. Some overlapped in time, however; and all began to develop teeth that were better suited to feeding on tough grasses, until the modern horse, known by scientists as *Equus*, first appeared.

World of Mesohippus

- Mesohippus was part of an evolutionary line of the ungulates. These are hoofed herbivores, such as cattle, deer, elephants, and horses.
- Mesohippus had adapted for life in the open grasslands. It had a longer snout than its predecessor. Hyracotherium, and was taller.

LIVED: in Oligocene times
SIZE: like a large dog
WEIGHT: unknown
DISCOVERED: in North America

OTHER DATA: herbivore; an ungulate; name means "intermediate horse;" ancestor of today's horses; three toes on front and back feet; lived on grassy plains; status: extinct

Andrewsarchus

In the American Museum of Natural History, New York, there are many outstanding exhibits, but one is quite remarkable. It is the the largest known skull of any carnivore, living or extinct. It is 33ins (84cm) long and 22ins (56cm) wide, and belongs to a meat-eater that died out 34 million years ago – Andrewsarchus, shown here as it prepares to attack.

LIVED: in Late Eocene times
SIZE: probably about 18ft (5.5m) long
WEIGHT: unknown
DISCOVERED: in Mongolia

OTHER DATA: named after Roy Chapman Andrews, leader of 1923 expedition on which first found; largest known creodont; say: AN-DROO-SARK-OOS; status: extinct

Dating from Late Eocene times and found in Mongolia, the fossilized skull that scientists believe belonged to a creature which they named Andrewsarchus is a fascinating specimen. It is the partial remains of an animal that was probably about 18ft (5.8m) in total length. At this size, it would have been about four times as long as one of today's wolves, and double the length of a brown bear.

Andrewsarchus is classed by palaeontologists as a *mesonychid*, which means it was one of the first group of prehistoric mammals to become specialized meat-eaters.

STRANGE CONNECTION

But what perhaps is the most curious thing about this wolflike creature is that palaeontologists think it is likely to have been a relative, if admittedly a distant one, not of today's wolves but of whales. There are certainly the most amazing similarities between the skulls and teeth of *mesonychids* and those of the earliest whales discovered so far.

World of Andrewsarchus

- Scientists have drawn up descriptions of Andrewsarchus on the basis of one fossilized skull only.
- Andrewsarchus has a very flat skull and enormous incisor teeth, ideal for biting into its prey.
- *Mesdonychids,* such as Andrewsarchus, walked on their toes, and had long tails and sharp teeth.

This connection was first suggested back in 1937; but since then there have been many more discoveries of *mesonychids* that seem to support the theory. An odd transitional creature called Ambulocetus, for example, dug up in what is now Pakistan, had limbs that were very well adapted for swimming, although there is no sign of it having had a tail fluke like a true whale.

Diprotodon

Today's marsupials – the opossum, wallaby and koala, for example – are by no means giant in size. Yet there were once huge members of this family – among them Diprotodon – that lived in Australia.

Scientists first found the remains of Diprotodon in the muddy deposits of a large lake in which some of these creatures had drowned as they fed at the water's edge. These remains were remarkably well preserved.

Unlike placental mammals – a group to which humans belong – marsupials have pouches in which the young are carried until sufficiently mature to fend for themselves. These creatures were once found worldwide but finally became extinct outside South America and Australia, where some are still found today. All are small or of moderate size, and the most familiar is probably Australia's kangaroo.

Way back in time, however, some were enormous in comparison with the marsupials of today. Australia's giant kangaroo, for instance, grew to 10ft (3m) in height. But even though it was so tall, it is thought to have got about in exactly the same way as a kangaroo of today – by springing and then bounding along on its two back legs. But Diprotodon, as you can see in this illustration, looked nothing like any of the marsupials with which we are familiar in the 21st century.

FACTFILE

All of the giant marsupials of Australia seem to have been hunted to extinction by humans about 12,000 years ago.

LIVED: in Pleistocene times, one million years ago
SIZE: up to 10ft (over 3m)
WEIGHT: unknown
DISCOVERED: in Australia

OTHER DATA: probably not very intelligent as had a small brain; a marsupial; 20-in (50cm)-long skull; powerful jaws; giant teeth; status: extinct

From them we can tell that Diprotodon was definitely an herbivore. Its teeth, for example, were ideal for chomping at vegetation.

Diprotodon footprints have been discovered; and the skeleton of a baby Diprotodon was even found exactly where it lay within its mother's pouch. Some Diprotodon bones, meanwhile, have marks that show they were hunted by humans.

The main enemies of Diprotodon and other large marsupials were probably creatures such as Thylacoleo, a marsupial lion with teeth like blades. This beast would have had no hesitation in attacking and bringing down large and sturdy quarry.

Platybelodon

About 10 million years ago, in what is now Mongolia, at the edge of the Gobi Desert, there roamed a mastodon that we know as Platybelodon. The illustration *below* shows you what a bizarre-looking creature it was! The jaws extended into a shovel-shape; and in three lower part there were two giant incisors.

LIVED: in Miocene times
SIZE: 6ft (1.8m) long
WEIGHT: unknown but probably several tons
DISCOVERED: in the Gobi Desert

OTHER DATA: found by American palaeontologist Roy Chapman Andrews; known for its extraordinary large jaws and shovel-tusk; *say:* PLAT-EE-BEL-OH-DON; status: extinct

About 5ft (1.5m) tall to the shoulder and with a skull and lower jaw extending to a length of approximately 6ft (1.8m), Platybelodon was much like an elephant except that it did not have a trunk. Those huge jaws took the place of one.

Platybelodon became extinct towards the end of Miocene times, about 6,000,000 years ago, and there is no creature alive today with such an extraordinary form of mouth.

A famous palaeontologist, Roy Chapman Andrews, discovered some of these remarkable creatures while on an expedition to the Gobi Desert in Mongolia in 1930.

As he wrote describing the find:

"Just below the tents on a narrow promontory, we discovered many outcrops of bones. When the deposit was opened, the skulls, jaws and skeletal parts of baby mastodons far outnumbered all other animals. Evidently this had been a bog near the shore of a lake. Mother shovel-tusked mastodons (Platybelodons) with their babies had come here to drink or feed."

All the specimens found were babies. Chapman Andrews and his team even found part of the skull of an unborn Platybelodon that was still within its mother's pelvis.

DEATH BY DROWNING

Not far away, there were other fossilized Platybelodon remains, too. In all, the team worked for six whole weeks to unearth these fossils. Chapman Andrews soon gathered what must have happened all those millions of years ago. There was probably once a bay where they were working, he reckoned, at the edge of which there may have been a well of soft, sticky mud. This was covered by water, however, in which grew lots of the aquatic plants that Platybelodon loved to feed on. Enticed by these plants, several of the creatures went further and further into the water, finally becoming stuck in the muddy terrain. Before long, they drowned; and so did many others, lured by the prospect of a large meal of vegetation in this way, their bodies finally becoming fossilized.

World of Platybelodon

- Roy Chapman Andrews and his team discovered the remains of male, female, and baby Platybelodons in Mongolia.
- Platybelodon's shovel tusk was a far more useful tool in many ways than the trunks of other mastodons.

Indricotherium

It may be hard to believe but this bizarre-looking, long-necked animal was related to the rhinoceros, even though it does not look anything like one and did not even have a nose horn of any kind.

If you could go back as far as 30 million years in time and travel to parts of Asia, you might come across an Indricotherium just like the one shown in the illustration *opposite*. It was more than twice the height of today's elephants.

Since it was first unearthed by palaeontologists, various names have been given to this creature. At one time, for instance, it was known as Baluchitherium, after Baluchistan in Pakistan, the area where it was first discovered. The current name, Indricotherium, meanwhile, is derived from the ancient Greek for beast, *therium*, and a mythical animal from Russian folklore, the Indrik, traditionally lord of all animals. First unearthed by the English palaeontologist C. Forster in 1910, fossilized remains of Indricotherium.have also been found in Mongolia and China.

Luckily, a good number of fossilized bones were dug up, as well as the skull. If just the skull had been discovered, scientists might have drawn the wrong conclusion, assuming that this animal must have been much shorter than it actually was because the head was small in comparison with the rest of its body.

MONSTER MAMMAL

Some scientists even believe that Indricotherium may have been the largest land mammal ever to have existed. (Dinosaurs, remember, were not mammals but reptiles.) It stood about 18ft (5.4m) high to its shoulders and was about 27ft (over 8m) in length. Scientists have estimated from its remains that, when fleshed out, it would have weighed as much as three large elephants.

It had massive, three-toed limbs and, in spite of its great height and stout build, could probably have run at a fair speed on its long, slim legs.

Living in small herds, it would take leaves and twigs from the tallest trees just as a present-day giraffe will do, using conical incisor teeth to chomp up its food.

FACTFILE

Horned rhinos which were relatives of hornless Indricotherium did not exist until long after this huge creature first appeared.

LIVED: 30 million years ago
SIZE: 27ft (over 8m) long
WEIGHT: as much as 2 elephants
DISCOVERED: in Pakistan, in 1910

OTHER DATA: herbivore; lived in small herds; 3-toed limbs; long, slim legs; a distant ancestor of today's rhinos, even though no horns; *say*: IN-DRI-KOH-THERE-EE-UM; status: extinct

Sometimes described as a "super-giraffe" rather than a relative of the rhino, Indricotherium would have towered above all other known creatures of its era.

Litopterns

No creatures alive today look as curious as some of the litopterns of prehistoric times. All were plant-eaters, and some resembled rabbits, horses or camels. The litopterns in this picture clearly had the most bizarre mouthparts.

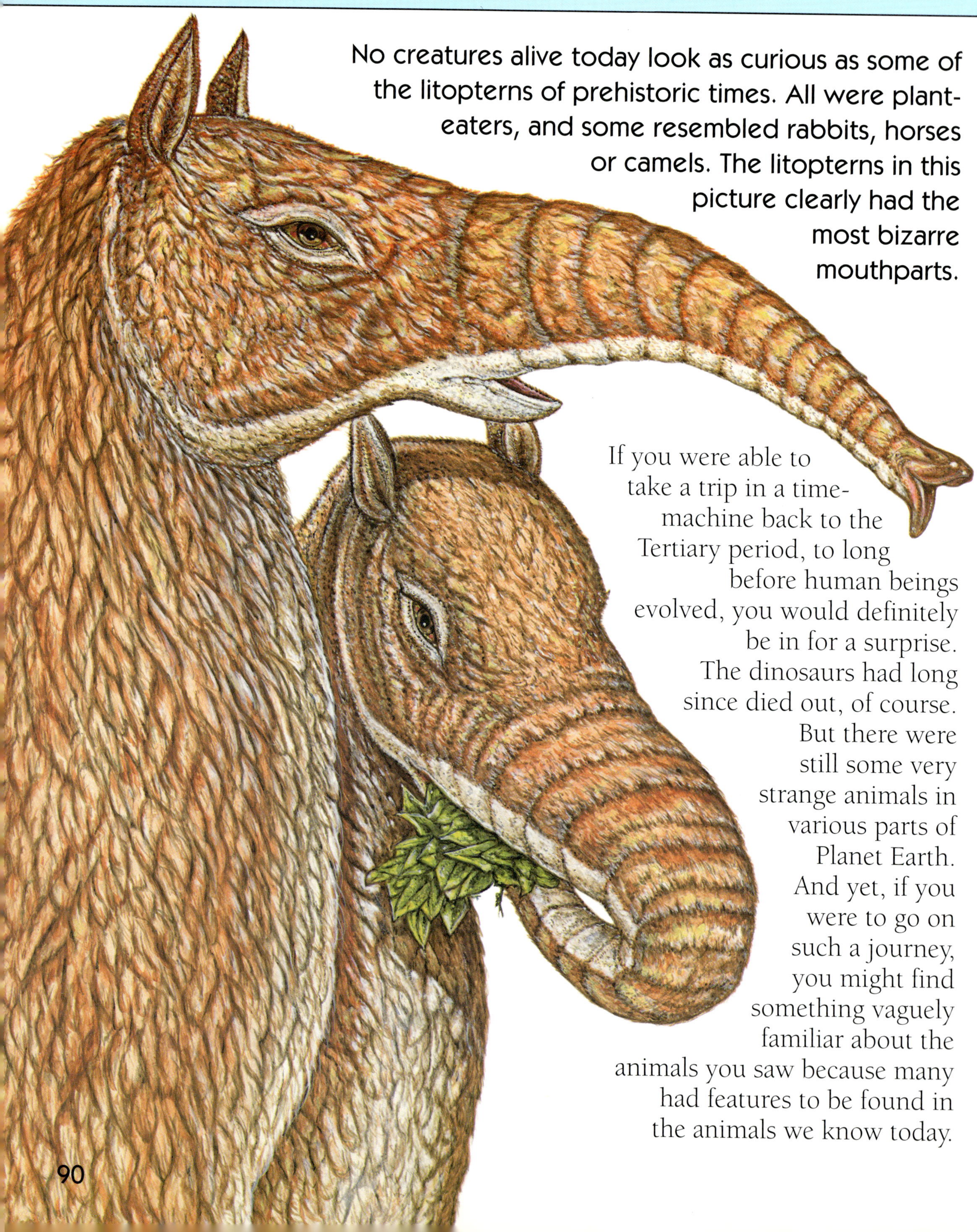

If you were able to take a trip in a time-machine back to the Tertiary period, to long before human beings evolved, you would definitely be in for a surprise. The dinosaurs had long since died out, of course. But there were still some very strange animals in various parts of Planet Earth. And yet, if you were to go on such a journey, you might find something vaguely familiar about the animals you saw because many had features to be found in the animals we know today.

LIVED: in the Tertiary period
SIZE: similar to a camel
WEIGHT: unknown but probably as much as a camel
DISCOVERED: in South America

OTHER DATA: herbivores; some had mouthparts resembling an elephant's trunk; partial skeletal remains found by Charles Darwin; soft body covering; status: extinct

During the Tertiary period, for example, there were two types of herbivores that were found only in that part of the world we now know as South America. One of these groups is known as the litopterns; the other, as the notoungulates.

Macrauchenia, two of which are shown in the picture *left*, were litopterns. Palaeontologists think that they had a body much like that of a camel except, that is, for a weird appendage like an elephant's trunk. This was no doubt very useful for gathering foliage from the tall trees on which they would feed. They may even have been able to spray water through it.

EXCITING DISCOVERY

The chances of being able to go back in time to see animals of the past are very remote. After all, no one has yet invented a functional time-machine, so it is for the moment the stuff of science fiction. Yet whenever palaeontologists unearth prehistoric remains, they feel something of the excitement of such an encounter. This is how it was for the great British naturalist Charles Darwin when he came across the partial skeleton of a litoptern. As he writes in his book entitled *The Voyage of the Beagle:*

"At Port Julian, in the red mud capping the gravel on the 90ft plain, I found half the skeleton of the Macrauchenia ... a remarkable quadruped, full as large as a camel. It belongs to the same division of the Pachydermata with the rhinoceros, tapir, and palaeotherium; but in the structure of the bones of its long neck it shows a clear relation to the camel, or rather to the guanaco ... I was at first much surprised how a large quadruped could so lately have subsisted in latitude 49 degrees .15, on these wretched gravel plains, with this stunted vegetation, but the relationship of the Macrauchenia to the guanaco, now an inhabitant of the most sterile parts, partly explains this difficulty." What a terrific find!

FACTFILE

Charles Darwin found and described the remains of several unusual prehistoric creatures during his 19th-century travels.

World of the litopterns

- No one is quite sure about the colour of the body covering of a litoptern such as Macrauchenia, but it was probably soft and silky, or maybe even woolly, rather than tough like crocodile or elephant hide.
- There were no doubt predatory carnivores that liked to make a meal of litopterns if given half a chance.

Brontotherium

Does the two-pronged horn that a Brontotherium had at the end of its snout remind you of anything? It was covered with skin and looked very much like a catapult.

What a massive beast Brontotherium was! It resembled the rhinoceros that we know today, but stood at about twice its height and had an even bulkier body.

Huge numbers of these extraordinary animals once roamed North America and eastern Asia. Just imagine what it would have been like to come across a whole herd of them coming in your direction! However, you would probably have been safe if you kept very still. They would not have attacked unless provoked. Herbivores, they lived on grass and other soft vegetation, and so had no need to be predatory and hunt for meat.

Part of a group known as *brontotheres* which first appeared during the early Eocene period – that is, about 55 million years ago, a considerable time after dinosaurs had become extinct – in turn, they then disappeared around 40 million years ago, in the middle of Oligocene times. Rhinos, more like those we know today, then evolved.

Indian, Javan and African rhinos all have horns, but some species have only one, while others have two or a few of which the front one is always larger.

UNUSUAL HORN

But no known surviving member of the rhino family has a single horn that splits to form two prongs like the one that Brontotherium had.

This peculiar protuberance is thought to have been used when fighting with rivals over females with whom to mate. The Brontotherium with the biggest divided horn may also have led its herd.

Classed as a *perissodactyl* (which means "having an odd number of toes"), four-footed Brontotherium had five toes on its front feet but only three toes on the back feet.

Its eyes were set towards the front of its head, and it had large nostrils which no doubt provided it with a good sense of smell. This would have helped it to sniff out pools among the grasslands of the plains. Here, it would have quenched its thirst and refreshed itself.

FINAL DEMISE

Fossilized remains of Brontotherium have been found in many parts of North America, including California's Death Valley, Nevada, Nebraska, Wyoming, and North and South Dakota.

But why did these lumbering giants finally die out? No one knows for sure, but the most likely reason is a dramatic change in climate that occurred about 40 million years ago. Far drier conditions would have brought about a lack of suitable foodstuff.

LIVED: 55-40 million years ago
SIZE: twice the height of today's rhinos
WEIGHT: twice the weight of today's rhinos
DISCOVERED: in North America and eastern Asia

OTHER DATA: herbivore; single divided horn; an early ancestor of the rhino; a *perissodactyl*; very bulky body; five toes on front feet, three on back feet; status: extinct

According to Native American legend, when the Sioux tribe first discovered the remains of Brontotherium, they used their imagination to make up a story about it. It was, they said, a type of horse that would thunder down from the sky when there was a bad storm. They were not too far wrong because scientists now know that Brontotherium was, in fact, a distant relative not only of the rhino it closely resembled but also of tapirs and horses.

FACTFILE

Some types of rhinoceros are severely endangered today as numbers dwindle, among them the Sumatran rhino.

Mesosaurus

A typical mesosaur, Mesosaurus was a small reptile, sometimes measuring as little as 16ins (40cm) in length. It lived in Permian times. As you can see in this close-up illustration, within its mouth there were masses of long, sharp teeth. It also had a lengthy tail that helped it to excel as a swimmer.

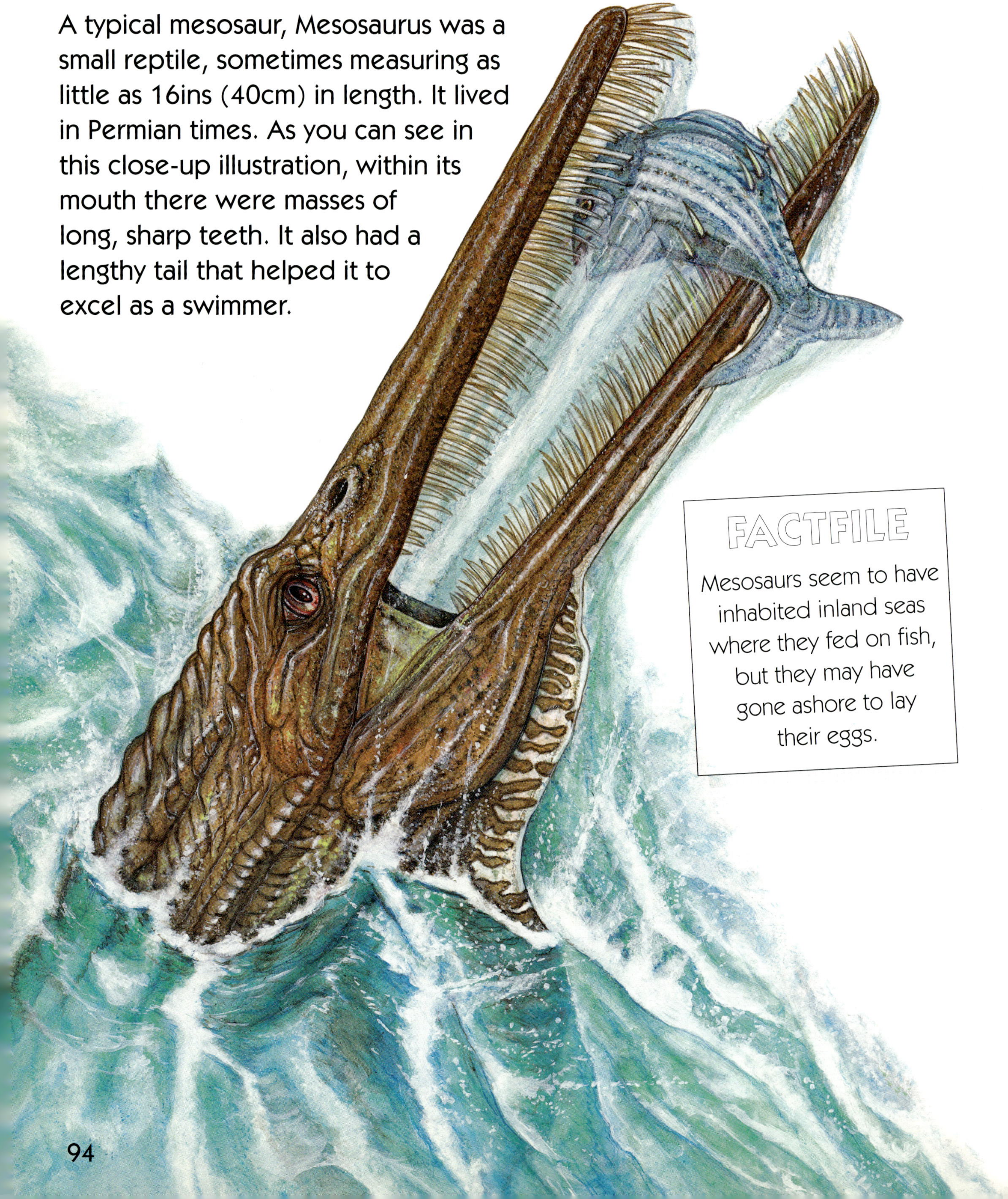

FACTFILE

Mesosaurs seem to have inhabited inland seas where they fed on fish, but they may have gone ashore to lay their eggs.

LIVED: before the dinosaurs, in Permian times
SIZE: 16ins (40.6cm) long
WEIGHT: unknown but not heavy
DISCOVERED: S. America, S. Africa, and Antarctica

OTHER DATA: reptile; lived in inland seas rather than oceans; many sharp teeth; possibly came on to land to lay eggs and give birth; long skull; flat-sided tail; status: extinct

World of Mesosaurus

- The very many long, brushlike teeth of Mesosaurus were ideal for catching slithery fish.
- Scientists think that Mesosaurus may well have had webbed feet which had a function similar to flippers in the water.

The fossilized bones of Mesosaurus have been found only in South America, South Africa, and Antarctica. It was, of course, very exciting that the discovery of a creature that lived 280 million years ago – long before the dinosaurs – had been made. But it did not stop there. From what has been unearthed so far by way of their remains, scientists have also been able to confirm a very important theory.

MOVING PLATES

When we look at an atlas today, it does perhaps become obvious that all the landmasses might vaguely fit together like the pieces of a giant jigsaw. But until a man called Alfred Wegener first suggested that the continents might have moved apart over millions of years, no one had had given much thought to the subject.

Wegener believed that, long ago, Europe and Africa had drifted apart from North and South America. But his idea was not well received at first because no one could understand what might have caused this.

We now know the answer, however. There is constant activity under the oceans, and the continents float on great sheets of the Earth's crust called tectonic plates. Back in Permian times, all land was linked together in one huge continent known as *Pangaea*, meaning "all Earth." Only in Triassic times and much later did this supercontinent start to break up very gradually into what scientists call *Laurasia* (comprising North America, Europe, and Asia), and *Gondwana* (comprising South America, India, Africa, Antarctica and Australia.) The two great continents then slowly, again over millions of years, broke up further until eventually the world took on the form that it has today.

By now you may be asking what this has to do with Mesosaurus! Well, the fact that its remains have been unearthed in both Africa and South America, and that it could not possibly have swum such a distance although perfectly at home in the environment of an inland sea, provides additional proof that these two continents were once part of the world's original landmass.

Deinogalerix

Discovered in rocks dating back to Late Miocene times in the south of Italy, Deinogalerix seems to have been an early form of hedgehog. But it is thought to have had a tail, and was hairy rather than spiny.

It may be that you have sometimes found a sweet-looking hedgehog in a garden, or perhaps you have come across one while walking in the countryside. If so, the chances are that it would have been curled up in a ball. Hedgehogs do this if they are at all frightened, as a form of camouflage in the attempt to protect themselves from danger. If it was convenient, you might even have offered it a saucer of milk if you suspected it could be hungry. But if you had existed when the hedgehog-like creature *below* was part of the fauna of this planet, would you have found it so cute?

Creatures resembling hedgehogs are thought to have lived as far back as Cretaceous times when dinosaurs such as Tyrannosaurus rex roamed the Earth. But Deinogalerix may not have evolved until far later.

FACTFILE

The name Deinogalerix has the meaning "terrible hedgehog." But no one knows if it was fiercer than the gentle hedgehogs of today.

LIVED: in Late Miocene times
SIZE: about 5 times bigger than today's hedgehogs
WEIGHT: about 5 times as much as today's hedgehogs
DISCOVERED: in southern Italy

OTHER DATA: hairy body; name means "terrible hedgehog;" first unearthed in 1973; when it existed, Italy was series of small islands; *say:* DEYE-NOH-GAL-ER-IKS; status: extinct

Its fossilized remains, first discovered in 1973, show it to have been several times the size of a modern hedgehog, both in height and length. In fact, it was probably the size of a small dog. You can get an idea of how large it was from the illustration *below*, in which it is sniffing at a prehistoric beetle. Other characteristics included a long, slim face; small pointed ears; a lengthy, tapering tail; and, of course, its long hairs.

Scientists believe it was an insectivore for the most part, probably feeding on beetles, dragonflies, and crickets; but it may also have hunted small mammals, reptiles, and birds, and could have enjoyed snails, too.

At that time, what is now Italy was a group of small islands and only later did most of these join with the mainland. It is, in fact, a known feature of islands that creatures isolated there sometimes grow to a different size than they would elsewhere. So Deinogalerix may have lived exclusively in that region. We cannot be certain unless its remains are uncovered in some other part of the world some time in the future.

Prehistoric survivors?

There are creatures alive today – turtles and crocodiles, for instance – whose ancestors were contemporaries of the dinosaurs. If these species have survived, could it possibly be that others also defied extinction to live on in remote, watery hide-outs in certain areas of Planet Earth?

World of the prehistoric survivors

- The word used for the study of strange, unidentified creatures that may be throwbacks is *cryptozoology*; and a scientist working in this field is known as a *cryptozoologist*. They need to spend a lot of time collecting reports about supposed sightings.

The waters of Lake Okanagan, in British Columbia, Canada, were chilly on the flesh, even though it was the height of summer. But what sent an even greater chill down the spine of the intrepid swimmer was another strange sensation.

DATA: sightings of creatures resembling ancient sea monsters sometimes reported; examples include Loch Ness Monster of Scotland and Ogopogo of Canada; such creatures seen in many other places, too; discovery of a live coelacanth fish, previously thought to have died out in prehistoric times, lends credence to stories of far larger surviving prehistoric beasts, but there is no concrete evidence

She felt she was being buffeted by a long, muscular shape which passed in a matter of seconds. Then, suddenly, some way ahead, there appeared above the surface three or four fleshy coils, each about 16ft (4.8m) across, which undulated unnervingly through the water of the lake.

If the swimmer is to be believed, this was a real-life encounter with a sea-dweller known locally for hundreds of years as Ogopogo. Several other sightings have also taken place, witnesses claiming that Ogopogo is grey in colour, with a snake-like body and whale-like flukes at the end of its tail. Other creatures of similar build have been sighted in lakes all over the world, such as Lake Van in Turkey and Lake Ikeda in Japan. Indeed, Japanese mothers to this day do not allow their children to play in the lake, for fear of meeting the creature that they call by the name of Issie.

Some people think that these supposed sightings are no more than ripples on the water made by the wind or passing boats; but others feel differently.

They claim that they are, indeed, sightings of ancient creatures. These, they say, did not die out entirely all those millions of years ago. Intriguingly, fossil remains of some prehistoric sea creatures do bear a marked resemblance to descriptions of today's elusive beasts. Several sightings on vast and lonely lakes have led cryptozoologists to believe that plesiosaurs are indeed possibly still alive and well. The "monsters" sighted in Loch Ness, Scotland, Lake Champlain on the Canada/US border, and Lake Khaiyr in eastern Siberia are all said to have a bulky body, an extremely long neck, and a small head. Witnesses cite as proof that these prehistoric sea creatures made it past a widespread extinction at the end of Cretaceous times, the remarkable similarity between the fossil evidence of a plesiosaur's flippers and the limbs they claim to have seen on these "modern" creatures. Their case is further strengthened by the discovery of the continued existence of the prehistoric fish, the coelacanth, once only known from its fossils.

FACTFILE

Scotland's Loch Ness monster is also said to be sighted every now and then, and murky photographs have been taken.

Extinct birds

Most of the birds in the section that follows have disappeared from the planet in comparatively recent times. The stories behind their extinction teach us many important lessons.

The chances are that most if not all of the birds illustrated across these two pages will be new to you. That, quite simply, is because they no longer exist. So the only place you would ever be able to see them would be in a museum collection as stuffed specimens or skeletal remains. So why have so many species of birds completely vanished? The very earliest disappeared along with the dinosaurs. But what about more recent extinctions?

In most instances, regrettably, humans have brought about the demise of these birds. The delightful dodo, for example, was hunted for its flesh on the island of Mauritius. Passenger pigeons were also shot for food, as was the great auk. And when some species became scarce, they were taken by collectors as specimens. Other birds, meanwhile, were hunted for their feathers for costume decoration or domestic use to fill quilts.

A number of species, however, were killed by farmers because the birds themselves were greedy predators of poultry or lambs and calves. The caracara, which you will meet on pages 128-129, would kill any creature weaker than itself even when it had other plentiful food supplies. Some birds have fallen victim to severe disease that wiped them out.

Natural disasters have also sometimes had a part to play in killing off a substantial proportion of species, which then went into dramatic decline. Some could once be seen in large numbers in the United States. Others come from places as far apart as New Zealand, the Solomon Islands, islands in the North Atlantic Ocean, and Africa. But they have perished.

Now turn the pages that follow and get acquainted with just some of the very many birds that so sadly no longer grace our skies.

Archaeopteryx

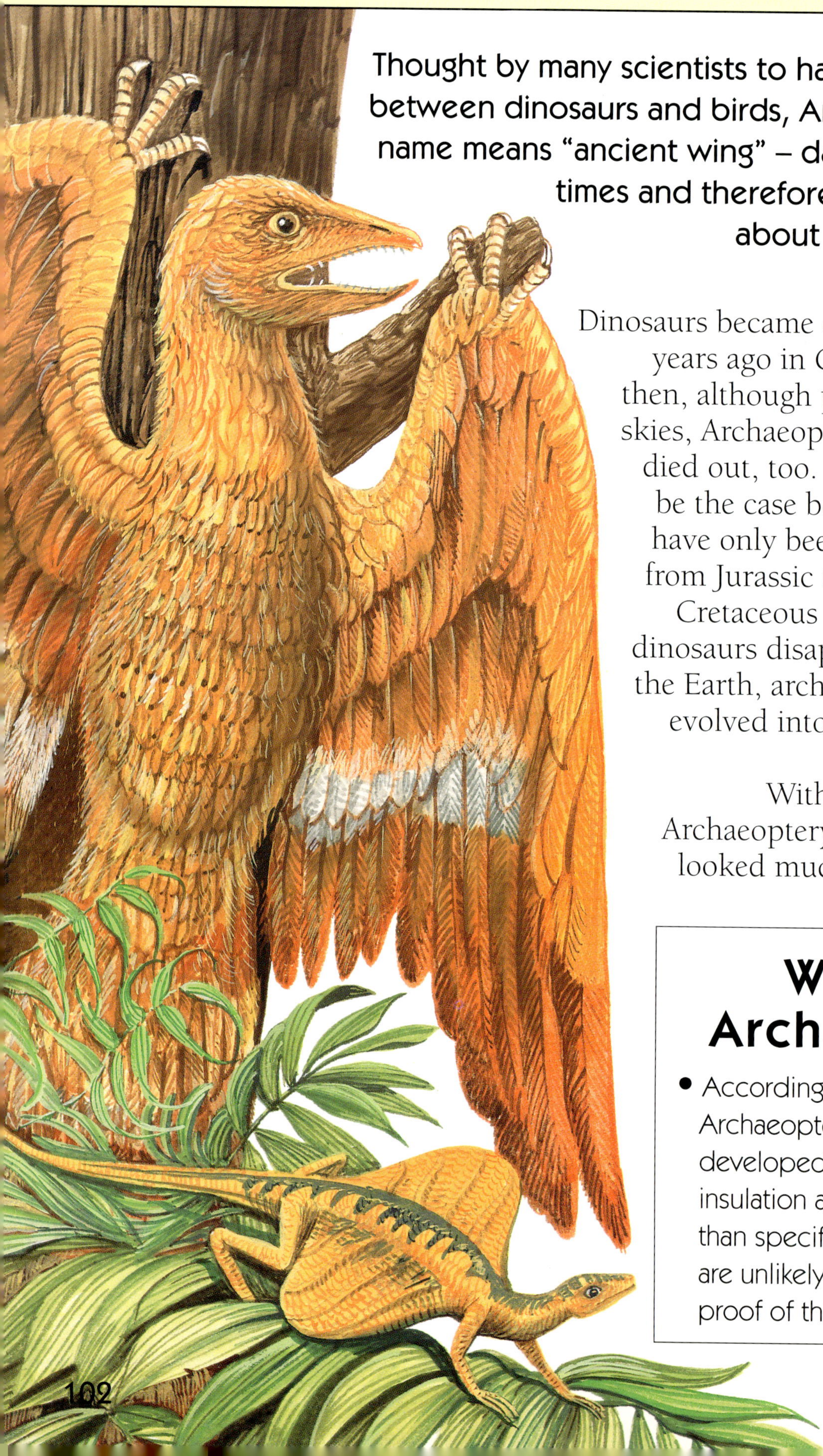

Thought by many scientists to have been a missing link between dinosaurs and birds, Archaeopteryx – whose name means "ancient wing" – dates from Late Jurassic times and therefore inhabited our planet about 155 million years ago.

Dinosaurs became extinct about 65 million years ago in Cretaceous times. But by then, although pterosaurs still ruled the skies, Archaeopteryx had probably long died out, too. Scientists believe this to be the case because so far its remains have only been found in rocks dating from Jurassic times which predate the Cretaceous period. By the time that dinosaurs disappeared from the face of the Earth, archaeopteryx had probably evolved into creatures more like the birds we know today.

Without its feathered wings, Archaeopteryx would probably have looked much like a small dinosaur.

World of Archaeopteryx

- According to some scientists, Archaeopteryx may first have developed feathers as a form of insulation against the cold rather than specifically for flight, but we are unlikely ever to have definite proof of this.

LIVED: in Jurassic time
SIZE: 15ins (38cm)
DISCOVERED: in Germany
DIET: insects, worms, lizards

OTHER DATA: fluttered rather than flying; had beak and toothed jaws; possibly the missing link between dinosaurs and birds; *say*: ARK-EE-OPT-ER-IKS; status: extinct

POOR FLIGHT

Its wings, however, were more like feathered arms, and Archaeopteryx was probably therefore more confident when it was climbing up trees or scampering along the ground than taking to the skies. We know that it could not fly well because it lacked the powerful pectoral (chest) muscles of today's birds, and it did not have a large furcula (wishbone) to aid flight.

When remains of Archaeopteryx were first found in a limestone quarry in Germany, a district medical officer who desperately wanted these fossils negotiated with the miners who had found them and offered free treatment in return. He then tried to sell the remains for a huge amount of money, and they were finally acquired by the British Museum in London, England, together with numerous other fossilized items he had collected over the years. Today, the price has increased many times over as Archaeopteryx fossils remain rare.

Fossilized feather prints have also been found. At first these were thought to have been made artificially as a practical joke, but they are now generally accepted as genuine and can be seen at the British Museum of Natural History, too.

FACTFILE

The only surviving bird with claws at the end of its wings like those of Archaeopteryx is the hoatzin of South America.

The head of Archaeopteryx may have looked birdlike, but its jaws were lined with teeth. (Today's birds, of course, do not have teeth! Instead they have beaks or bills.) Its tail, meanwhile, was long and bony, again not at all like the tails of modern birds. Its size is thought to have ranged between that of a chicken and a small turkey. (Not all dinosaurs, remember, were large.) So in some respects, Archaeopteryx definitely seems to have been like a reptile; and, in others, more like a bird.

It possibly was indeed the missing link that some experts now believe it to have been.

FEW FOSSILS

Whatever the case, why have so few Archaeopteryx remains been found? One suggestion is that because their bones were so fragile, they broke up very easily and disintegrated completely over millions of years, except for the precious few that have been found.

Its feathers seem to have been like those of today's pigeons. In fact, the shape of these has suggested to some scientists, who doubt the theory that feathers originally evolved purely as a form of protection against the elements, that they were primarily an aid to flight right from the start.

Sinosauropteryx

The discovery in a remote corner of China of fossils that looked like dinosaurs with feathers may be final proof that dinosaurs are in fact still with us – but in the form of birds!

It is not hard to imagine that a small, furry Megazostrodon knew something might be amiss as it scampered along speedily. It had suddenly found itself by the side of the long tail of a potential predator which, though only chicken-sized, would almost certainly move in for the kill.

With its strong sense of smell, the Sinosauropteryx was already well aware of the approach of its prey. Now all it required was a quick flex of the short, sinewy neck of the hungry birdlike reptile and the unfortunate tiny Megazostrodon would be clasped firmly in its sharp-toothed jaws, ready to be swallowed whole.

FACTFILE

The scientific name given to this early bird has the meaning "first Chinese dragon feather."

That Sinosauropteryx was an early ancestor of birds should no longer be in any doubt, as experts who have examined fossils that were dug up in Sihetun, north-east China, in 1996, have been quick to confirm.

Like its modern counterparts, the 120 million-year-old creature they discovered had hollow bones; three-toed, forward-facing feet; and a bird-like neck. But what was particularly remarkable was its covering of fine filaments, each up to 1.5ins (3.8cm) long. These, experts claim, were the forerunners of feathers. However the fibres of these filaments seem to have lain parallel and were not branched into barbs and barbules as flight feathers are. Perhaps, therefore, they were structures that would eventually evolve to become feathers. Indeed, it is now thought that feathers almost certainly evolved from reptile scales.

LIVED: in Jurassic times
SIZE: 20ins (51cm)
DISCOVERED: in China
DIET: worms, insects, lizards, small mammals

OTHER DATA: scientific name, *Sinosauropteryx prima; say*: SEYE-NOH-SAW-OPT-ER-IKSslimly built; early ancestor of birds; long filaments instead of feathers; flightless; status: extinct

DINOSAUR-LIKE

Sinosauropteryx also seems to have looked remarkably like such lightweight dinosaurs as Compsognathus.

With slim bones and fleet of foot, they shared a similar skeletal structure, particularly in their arms, shoulders and chest. This finding also mirrors the discovery in Germany, in 1861, of fascinating Archaeopteryx remains, as described earlier on pages 102-3.

Dodo

The islands of Mauritius and Réunion in the Indian Ocean were once home to the dodo, a most delightful, flightless, large-beaked bird. But it was hunted to extinction by both humans and other animals.

No one is quite sure how the dodo got the name by which it eventually became known. But one possibility is that it was taken from a word meaning "idiot" – *duodu* (pronounced as DOO-OH-DOH) – by Dutch sailors.

Perhaps they thought that because this bird could not fly, it was stupid. Certainly, it could not easily run very fast, as its large belly almost scraped along the ground. Or maybe it was named after its call. (Try saying "*dodo*" through your nose, and see if you can make it sound like a bird call.)

BULBOUS BEAK

There are several written accounts from the 17th century, when it became extinct, that describe this bird. As you can see in this illustration, its main feature was a large, bulbous beak – probably its main form of defence since it had to stay and face an enemy rather than flying off.

LIVED: prior to end 17th century
SIZE: 3.25ft (1m)
DISCOVERED: on the island of Mauritius
DIET: fruit of the dodo tree, and perhaps insects

OTHER DATA: scientific name, *Raphus cucullatus*; *say*: RAY-FOOS-KOOK-OO-LAH-TOOS; featured in Lewis Carroll's book *Alice in Wonderland*; strange beak; flightless; status: extinct

Indeed, its bite is likely to have been quite painful to a victim.

The dodo's eyes were large and have been described as being like diamonds and even swollen gooseberries!

For the most part, dodos lived on their own and were solitary creatures until the breeding season. Then they mated for life.

Contemporary accounts describe these birds as being excellent parents, both males and females doing everything to protect their young.

FACTFILE

Legend has it that two Réunion dodos, shipped to France, actually shed tears of sadness when they had to leave the island.

Passenger pigeon

One hundred and fifty years ago, between five and ten billion passenger pigeons flew the skies over North America, but now there are none. Why did they disappear?

The sight of a flock of passenger pigeons overhead is said to have been absolutely amazing.In fact, when the eminent American ornithologist and artist John James Audubon reported seeing such a flock, he wrote that: *"The light of the noonday sun was obscured as by an eclipse."*

Whenever they spotted what looked like a good feeding area, the pigeons would all swoop down and circle in a formation that looked much like a rolling cylinder.

On average, a single passenger pigeon probably consumed the equivalent of one cup of seeds, berries, fruits, nuts, worms and insects every day; and if they were ever tempted by something else that was particularly tasty, they would regurgitate the last meal.

World of the passenger pigeon

- Flocks of passenger pigeons made a deafening racket at times and extended as far as 320 miles (514km.)

The passenger pigeon lived only in the United States, and most inhabited areas of deciduous forest, ranging as far north as the border with Canada and down to the southern states. Everywhere these birds went they would leave their mark, and the ground would be full of their droppings. Tree branches were also often broken in half because of the sheer weight of their numbers.

Several theories have been put forward to explain the disappearance of this species. Some experts suggest that disease was to blame; but most now agree that it can be put down to the hunting of this bird for human food. Passenger pigeon pie was certainly once frequently on the menu in many American households. These birds were also cruelly treated. At times they were trapped and then kept alive, only to be released at events where they would be shot for sport. The last passenger pigeon, named Martha by its keeper, died in 1914 in Cincinnati Zoo.

LIVED: prior to 1914
SIZE: 16ins (40cm)
DISCOVERED: in North America
DIET: acorns, nuts, fruit

OTHER DATA: scientific name, *Columba migratoria*; *say*: KOL-UM-BEE-AH MEYE-GRAT-OR-EE-AH; flew in huge flocks in swirling formation; most numerous bird ever; status: extinct

Choiseul pigeon

On Choiseul, one of the Solomon Islands which lie not too far from New Guinea, a great English naturalist, A.S. Meek, identified a large and very attractive crested pigeon. In 1904, it was given the scientific name of *Microgoura meeki* in his honour.

FACTFILE

The last sighting of a Choiseul crested pigeon is thought to have been in the Second World War; but there is no definite proof of this.

If it were not for the naturalists sent by English collector Walter Rothschild to exotic places in order to obtain samples of the animal life there for his remarkable collection, no one in the West might ever have heard about the Choiseul pigeon.

A.S. Meek was one of these collectors and he must have been extremely brave because venturing to very remote regions was fraught with dangers. As he put it:

"The natives of Choiseul are not only savage to strangers, but are by no means friendly among themselves."

In fact, the natives were such a threat that not all Meek's party would go ashore at one time so that they could rush to assist if necessary.

LIVED: prior to mid 19th century
SIZE: 12ins (30cm) long
DISCOVERED: on Choiseul Island
DIET: probably nuts, fruit, seeds

OTHER DATA: scientific name, *Microgoura meeki*; low, trilling call; fine head crest; never found in large numbers; may have been killed off by cats; *say:* SHWA-ZERL; status: extinct

Meek found the skins of six specimens and also a cream-coloured egg. But what was the pigeon itself like? The beautiful bird had a black face that was brightened by a reddish hue, a blue crest, and purple legs. It lived mostly in the swampy forests of the island, roosting on low branches. Only rarely flying in the open, it was principally land-dwelling, and it was this that probably led to its unhappy demise.

At one time, cats were brought to the island, specifically to help cut down the enormous rat population. But these cats also became predatory to the Choiseul pigeons, which had slow reactions and so frequently fell victim to the greedy felines, as shown *below*.

World of the Choiseul pigeon

- The natives of the island of Choiseul called this lovely low-flying bird the *kukuru-ni-lua*. In translation, this means "bird belong ground."
- The natives were particularly skilled at imitating the trilling of these pigeons.

Native islanders explained that the Choiseul pigeon had never been very common; but another factor which may have led to the complete disappearance of this pigeon on the island was the virtual elimination of its natural habitat. This happened when coconut plantations were introduced to replace swampy ground for economic reasons. Researchers hope to find this pigeon still inhabiting other nearby islands, but so far there has been no success.

Great auk

There were once millions of great auks living in the waters of the North Atlantic Ocean and along its coastlands. But the very last pair to have survived was sighted in 1844. This species had a sad demise.

In 1996, British author Errol Fuller, who has written a splendid book all about the great auk, splashed out and spent his entire life savings on a magnificent stuffed specimen of this bird. He bought it from a French baron, in whose home it had been gathering dust for nearly 150 years.

No wonder Fuller has become so fascinated by these extinct birds! Also known as garefowls, they looked as cute as penguins but had large, striped beaks, and there was a similarity in their behaviour – they, too, could not fly. But there was an important difference. Penguins, of course, all live *south* of the equator in Antarctic regions, whereas great auks were only found in the *northern* hemisphere.

The great auk's wings were tiny and useless for taking to the air. On land, it could only lumber along on its webbed feet; so if chased by a predator, it did not stand a chance. But once this bird took to the water, its wings would be used as flippers.

World of the great auk

- Fossilized remains of the great auk have been found in the United States as far south as Florida. So long before its habitat was restricted to northern waters, it must have lived this far south.

BREEDING GROUNDS

Great auks were hunted and cruelly clubbed to death for centuries by people of the far North who sought their flesh and feathers. It is even sometimes said that it was the migration of the great auk to breeding grounds at certain seasons of the year that first encouraged the Vikings to try to find new shores by sailing westwards across the Atlantic to the eastern shores of North America about one thousand years ago.

LIVED: prior to 1844
SIZE: 30ins (76cm)
DISCOVERED: in the North Atlantic
DIET: mainly fish

OTHER DATA: scientific name, *Alca impennis*; *say*: AL-CA IMP-EN-IS/ORK; also called the garefowl; killed for its meat and feathers; huge beak; clumsy on land; status: extinct

The killing of the last surviving pair occurred on the island of Eldey in 1844. Three men then sold the bodies to an Icelandic collector.

Giant woodpecker

The American ivory-billed woodpecker was so strong and determined that, as depicted in this illustration, one ornithologist reported it capable of pecking its way through ceiling plaster in the attempt to make an escape and fly free to the outside world once more.

LIVED: prior to 1972
SIZE: 20ins (60cm)
DISCOVERED: in North America
DIET: insects, worms

OTHER DATA: scientific name, *Campephilyus principalis*; *say*: KAMP-EF-EEL-OOS PRINS-EE-PAL-IS; logging destroyed its habitat; red head crest; status: extinct

Ornithologist Alexander Wilson once reported that he had wounded an American ivory-billed woodpecker, one of the giants of this type of bird. Having left it to recover in the shelter of his hotel room, he was surprised on his return after just a short absence to find it was not only up and about but energetically trying to make a hole high up in the wall. One moment it had been lying there as if half-dead; the next, it was full of life, clearly making a determined attempt to escape from the room. If Wilson had not disturbed the bird in this task, he was sure it would have managed to succeed in getting out.

HANDSOME BIRD

From a study that was commissioned by the National Audubon Society in 1942, we know that the American ivory-billed woodpecker was about 20ins (50cm) in length and that it got its name from its cream-coloured beak.

Never found in huge numbers, it was a very attractive bird with a bright red head crest, and found mostly in the forested regions of the south-eastern United States. Indeed, it was probably the destruction of such areas for arable farmland that was to lead to the bird's extinction.

By the late 1930s, it has been estimated that there were probably under twenty-five of these birds in existence. But there were a few sightings after that – even as late as 1969 and in the state of Florida.

CROWNING GLORY

The ivory bird itself was so admired that it became a prized item among certain Native Americans, particularly some of those living in Canada.

In his 18th-century work, the *Natural History of Carolina*, Mark Catesby wrote:

"The bills of these birds are much valued by the Canada Indians who make coronets of them for their princes and great warriors."

Apparently, they would exchange a single bill for two or three buckskins. So the killing of the birds for this purpose may also have contributed to their demise.

Ornithologists have come up with the disappointing news that the Cuban variety of this woodpecker may also now be extinct. A male was seen in flight in 1968; but since then there have been no confirmed sightings. We can only hope that there are indeed live specimens of both varieties that have chosen to lie low for the moment.

FACTFILE

The American ivory-billed woodpecker finally became so rare that it was considered extinct by 1972.

Spectacled bird

Discovered by Georg Wilhelm Steller, the eminent 18th-century German naturalist on Bering Island, the spectacled cormorant was given this common name because of the distinctive pale markings round its eyes. These, at first glance, gave the impression that it might have been wearing a pair of glasses!

In 1741, the Danish explorer Vitus Bering set sail in his ship, the *St Peter*, on a great voyage of discovery. The aim of his expedition was to find out whether there might be a land bridge linking Asia and North America. He succeeded in reaching the coast of Alaska, and was honoured for his great efforts.

LIVED: prior to mid 19th century
SIZE: 38ins (96.5cm)
DISCOVERED: on Bering Island
DIET: mainly fish

OTHER DATA: scientific name, *Phalacrocorax perspicillatus; say:* FAL-AK-ROK-OR-AKS PER-spik-il-aht-oos; eye markings resembling spectacles; status: extinct

The stretch of water that the *St Peter* had crossed, with its companion ship, the *St Paul*, was named after him and is known today as the Bering Sea.

The expedition was a fascinating one; but on the return journey the crew was shipwrecked in a dreadful storm and found themselves on an unhabited island, now also named after Bering.

Among the crew was Georg Steller, who as well as being a naturalist also doubled as the ship's doctor. Sadly, he did not manage to save Bering's life; but he did find some wonderful new species on the island. Among them was the spectacled cormorant.

According to Steller, these birds were to be found in huge numbers on the island. They were slow-moving and poor fliers, however. This meant that they were easily caught. Weighing as much as 14 pounds (6kg) apiece and cooked, unplucked, in clay, they could provide a substantial and highly nutritious meal for a lot of people.

The female birds were similar in size and shape than the males. However, the females did not deserve their popular name because their markings were different. They had no distinctive crest and no thick skin forming "spectacles" around their eyes.

FACTFILE

In addition to the spectacled cormorant, German naturalist Georg Steller also discovered a sea cow which was named after him.

Over the years, Bering Island came to be visited as a stopping-off point by whalers and other hunters of marine creatures, and was eventually settled by immigrant peoples.

FOOD FOR THOUGHT

They, too, must have enjoyed feeding on succulent spectacled cormorant meat; because by the middle of the 19th century – just 100 years after Steller had first come across them – spectacled cormorants (known scientifically as *Phalacrocorax perspicillatus*) were no more.

Later it was found that a few had managed to struggle to a nearby island. But here, too, they were frequently slaughtered for food.

Today there are only a few stuffed specimens that remain. These can be seen in the British Museum of Natural History, London, and museums in St Petersburg and Helsinki, capital cities of Russia and Finland respectively.

World of the spectacled bird

- The spectacled cormorant had a curious s-shaped neck which it could twist this way and that very easily.

Laughing owl

When in flight, just before a downpour, or when it was actually raining, laughing owls would suddenly produce a whole range of curious cackles and cries.

Native to New Zealand, the whekau became extinct at the beginning of the 20th century. Known by the nickname "laughing owl" because of its range of sounds, this bird would only come out into the open during the day if disturbed, and even then it always gave the impression that it was sleepy. In general, like most owls, it would remain in its hiding place with its eyes half-closed until dusk.

The laughing owl was the largest of New Zealand's owls and had a prominent beak and long legs, but weak wings, making it a good ground-hunter. Its plumage varied in shade. Some birds were darker than others, and some naturalists believe they may even have changed colour according to the season or at different stages of their lifecycle.

World of the laughing owl

- Female laughing owls usually laid two white eggs each season, and both parents shared the duty of incubation.

The whekau would make its home in the crevices of a rock outcrop or in caves. In general, the adults fed on rats, beetles and small reptiles; and their extinction is sometimes linked with the decline of the Maori rat, which had been brought to New Zealand by the Polynesians centuries before this bird disappeared. But other mammals may also have contributed to the decline of the whekau. Ferrets and weasels had been introduced in order to decrease the huge rabbit population, and these may also have been predatory to the owls.The last confirmed sighting of New Zealand's laughing owl was in 1907, almost one hundred years ago.

Competition is sometimes very keen among collectors of specimens of extinct birds, and high prices are frequently paid. During the 19th century, specimens of laughing owls were sent to Great Britain, and many fetched several thousands of pounds.

The story is even told about a confidence trick that was once played by an expert who was trying to fool another. An eminent British banker and naturalist, Walter Rothschild, was offered a supposedly new species of laughing owl, but he soon realised that the tail came from another type of owl altogether and that the specimen was therefore not genuine. The tail had merely been stuck in position. No one knows where this fake is today.

LIVED: prior to 1917
SIZE: 19ins (48cm)
DISCOVERED: in New Zealand
DIET: worms, beetles, rats, lizards

OTHER DATA: scientific name, *Sceloglaux albifacies;* *say:* SKEL-OG-LOHKS ALB-EE-FAK-EE-AYS disliked daylight; a cry like laughter when flying, hence its name; status: extinct

Mysterious starling

Sometimes, in the past, interesting specimens of rare creatures were left to decay simply because no one knew how to preserve them properly. One such bird was the appropriately named mysterious starling.

As if it was not disappointing enough that a stuffed item in the collection of the British Museum of Natural History, nicknamed the mysterious starling and known scientifically as *Aplonis mavornata*, was so badly preserved, over the years it became covered in dust and was completely neglected. Soon it deteriorated so badly through exposure to light that all remaining of this extinct bird was its skin.

No one, however, was entirely sure where it had originally come from until quite recently.

For a long while, it was assumed that it had been brought back from some Pacific island after the third of Captain Cook's voyages in the 18th century.

On this particular voyage, the surgeon acting as naturalist fell ill and died; so the rest of the crew was left to handle all the specimens of animal and plant life that had been collected. None would have been sufficiently skilled to do this correctly, however, and probably had no idea of how important many of the items were.

Indeed, none of the bird specimens thought to have been brought back to England after Cook's third voyage has survived – apart, that is, from the skin of the mysterious starling.

In 1986, however, an article was published in New Zealand by an expert on extinct birds living in America. Its author had studied a previously unknown manuscript, also in the British Museum, which seemed to describe the very same bird.

The manuscript had been written by the naturalist Andrew Bloxham who had sailed to and from Hawaii in 1824 to return the corpses of the unfortunate King and Queen of Hawaii to the island of Honolulu. (They had died while on a visit to England as the result of a virulent attack of measles.)

On the return voyage, the ship happened to call at the Cook Islands; and, as recorded in the manuscript, during the brief stay, Bloxham shot three birds there which he thought of interest. One of these he described as a pigeon, another as a kingfisher, and the third as a starling.

FACTFILE

The mysterious starling skin in the British Museum collection was first brought to light by the ornithologist Richard Bowdler Sharpe.

LIVED: prior to mid 19th century
SIZE: 7ins (18cm)
DISCOVERED: on the island of Mauke
DIET: probably fruit, nuts, seeds

OTHER DATA: scientific name, *Aplonis avornata*; *say*: AP-LEE-OH-NIS AV-OR-NAH-TAH; specimen in British Museum of Natural History; status: extinct

It is known that what was brought home from this voyage was presented to the British Museum, and so current thinking is that the surviving skin may not be from the time of Captain Cook after all, but rather a specimen brought back by Bloxham.

In 1970, an ornithologist visited the very island, Mauke, on which Bloxham had shot the starling but could find no trace of these birds. Could it therefore be that this very shooting of a rare bird finally led to it becoming extinct?

Whether or not the collecting of rare species in this way is a worthwhile idea provides lots of food for thought. What is *your* opinion?

Carolina parakeet

The only species of parrot with a natural habitat in the United States was treated with distain and often killed.

The illustration, shown *left*, has been based on a magnificent 19th-century painting of Carolina parakeets that can be found in John James Audubon's collection of ornithological paintings entitled *Birds of America.* This artist has certainly left us a magnificent legacy of images of birds, several of which are now extinct. But he also admitted to shooting a whole basketful of Carolina parakeets in order to get some specimens to use as reference for his paintings. Most people today would no doubt find it difficult to accept that so many birds, if any, though common at the time, had to be killed to produce his work. Audubon may have regarded them as pests, however, because enormous flocks of Carolina parakeets often destroyed all of a farmer's crops. Apples and pears would also be eaten by these birds before they had ripened, destroying whole orchards.

LIVED: prior to 1918
SIZE: 12ins (30cm)
DISCOVERED: in North America
DIET: in North America

OTHER DATA: scientific name, *Conuropsis carolinensis*; *say*: KON-YOUR-OP-SIS KAR-OL-EYE-NEN-SIS; bright yellow heads; shot due to its destruction of crops; status: extinct

Audubon even wrote that huge numbers sometimes covered the fields *"so entirely that they present to the eye the same effect as if a brilliantly colored carpet had been thrown over them."*

Carolina parakeets lived mostly in Florida, but there are claims that some were sighted flying as far north as the state of New York and even close to the Great Lakes of North America.

A DYING BREED

Carolina parakeets were first classified by the great Swedish naturalist Carolus Linnaeus in 1758, and were given the scientific name *Conuropsis carolinensis*. As with other parakeets that have been said to be extinct over the years, occasional sightings of surviving Carolina parakeets were claimed at first, but none has ever been substantiated.

But could there have been other reasons, apart from shooting by humans, for its final disappearance? Scientists think that it may have been due in part to the destruction of the forest environment that it had favoured.

THE FINAL PAIR

There were at one time several Carolina parakeets at America's Cincinnati Zoo but they did not breed well in captivity, and soon only two remained. Their names were Lady Jane and Incas.

The story goes that Great Britain's Zoological Society was so keen to acquire this rare pair that its administrators offered a very large sum of money for them, but it was not accepted. Cincinnati Zoo was not prepared to part with these birds.

LOST FOREVER

Sadly, Lady Jane died in the summer of 1917, and Incas no doubt pined for her terribly. He no longer had a companion with whom to share his cage and died himself only a few months later. His keepers claimed it was not as the result of old age, nor disease, but because he did not have the will to survive without her. It is hardly surprising. The world's very last pair of Carolina parakeets had lived together for 32 years.

World of the Carolina parakeet

- Carolina parakeets were gregarious by nature and would often rest in large groups, clinging to the bark of trees or even huddled together in a stump.
- Egg-theft will often lead to a bird's demise. Some of the Carolina parakeet's were stolen back in the 1920s by a curator at the University of Florida.

Paradise parrot

At the beginning of the last century, many people liked to keep songbirds as pets, and thousands of them were exported to Europe for this purpose. If this demand had not been so great, perhaps the highly vocal paradise parrot would have survived in its natural habitat.

They were never found in large numbers; but the very last surviving pair of paradise parrots is said to have been spotted in the state of Queensland, Australia, in 1927, about 80 years after they were first recorded by naturalist John Gilbert.

What lovely birds they were, to judge by reports dating from that time! The males had a red forehead, and lower parts of their abdomens and thighs were red, too. The females, however, had white-yellow foreheads and pale blue abdomens, sometimes flecked with red.

They were very tame, and so it would have been quite an easy task to catch one, which would only have led to a decrease in numbers in the wild.

But there was something bizarre about their behaviour. Instead of nesting in trees or on the ground like most other birds, the female paradise parrot usually burrowed into a termite mound.

She would work hard at pecking out a tunnel into the structure of the termite mound, and this would lead to a chamber that was sufficiently wide for herself and the three eggs that she would lay. after mating.

The male, however, was never a welcome visitor to the chamber that his mate occupied inside the mound. Instead, he would stand on or near the mound and call to the female from outside.

FACTFILE

The second word of the scientific name for the paradise parrot, *Psephotus pulcherrimus*, means "very beautiful."

LIVED: prior to 1927
SIZE: 11ins (28cm) long
DISCOVERED: in Australia
DIET: grass seed mostly

OTHER DATA: scientific name, *Psephotus pulcherrimus; say:* SEF-OH-TOOS PUL-KER-EE-MOOS; very tame; females nested in termite mounds; liked to feed on grass seed; status: extinct

DISASTER STRIKES

But capture by humans and a caged life as a prized pet were not the only problems to beset this species. New types of crops were gradually being introduced to what had formerly been a natural habitat of open grasslands. This greatly helped the economy of the region, and farmers prospered. However, there was a downside, and all this agricultural activity began to harm much of the local wildlife. The paradise parrot in particular was affected as its food supplies quickly became more and more depleted.

Whereas once these parrots had been seen feeding to their hearts' content on grass seeds as they balanced precariously to feast directly from the stalks, now they could only scavenge for scraps. And there was worse to come. Drought occurred so that plant life suffered, and then a series of bush fires took its toll. Meanwhile, cats, dogs and foxes all probably fed on paradise parrots whenever they got the chance.

In 1918, there was a newspaper campaign to try and find any surviving paradise parrots, but not until four years later was there any response.

Someone had seen a pair!

He watched them over several months, taking photographs to prove the sightings; but they, too, eventually vanished. The animated song of the paradise parrot may never be heard again.

Mamo

The cloaks of traditional Hawaiian chiefs were usually decorated with feathers; and, in some instances, these came from the very attractive mamo – a bird first named by ornithologists in 1789.

One of the most well-known Hawaiian ceremonial cloaks is so magnificent that experts have estimated it must have taken the feathers of as many as 80,000 mamos to embellish it!

Those who were given the task of catching the birds were supposed to set them free after the feathers had been plucked; but how long any of them would have been able to survive without the protection of their plumage is not known. Presumably, they soon perished.

The mamo probably became extinct at the very beginning of the 20th century as the result of such intensive hunting, but changes to its natural habitat may also have been at least partly to blame.

The call of the mamo was said to be a single, plaintive note, and it fed mostly on nectar and small insects. One collector described a mamo he had caught in a snare to take back to England as a specimen as follows:

"He is a beauty and takes sugar and water eagerly, and roosts on a stick in the tent."

LIVED: prior to late 19th century
SIZE: 8ins (20cm)
DISCOVERED: in Hawaii
DIET: mostly nectar and insects

OTHER DATA: scientific name, *Drepanis pacifica; say,* DREP-AH-NIS PAS-IF-EEK-AH; mournful song; beautiful plumage; long, curved beaks; status: extinct

Adult mamos grew to about 8ins (20cm) in length and were a glossy black. Parts of their plumage were a bright yellow, however. Their beaks, like those of most nectar-lovers, were long and very curved, allowing them entry into plants and to snap up insects in flight.

In 1898, one collector succeeded in locating a small group of mamos flitting about in some woods in Hawaii and took a pot shot at one. First it fell from the branch but just managed to recover itself before reaching the ground and flew to the other side of the tree on which it had been resting. Here it was joined by another of its species, perhaps a parent or a mate.

Luckily for this bird, it could not have been too badly injured because the pair completely disappeared, the collector assuming they had flown off. The noise of his shot must also have scared the rest of the group as they, too, vanished from sight. None has been seen since.

FACTFILE

A less glossy species of mamo, known as the black mamo, also inhabited Hawaii but is now thought extinct, too.

Caracara

A ferocious predator of poultry and other small or young domesticated animals, the caracara seemed to enjoy killing creatures weaker than itself even when food was abundant.

LIVED: prior to late 19th century
SIZE: 22ins (55.8cm)
DISCOVERED: in Guadalupe
DIET: shellfish and insects mainly

OTHER DATA: scientific name, *Polyborus lutosus*; *say*: POL-EE-BOR-OOS LOOT-OH-SOOS; preyed on poultry and newborn goats; loved to kill; silent unless attacking; status: extinct

World of the caracara

- Guadalupe caracaras would callously kill one of their own kind if one of them happened to become ill or disabled.
- The Guadalupe caracara was similar to the common caracara found on the mainland of California but far fiercer.

It must have been horrific to witness a Guadalupe caracara tearing its prey apart. It would even pull out a victim's tongue if given the chance in the attempt to incapacitate it.

Among its favourite prey were young goats. A 19th-century writer, who witnessed caracaras attacking these newborn creatures on many occasions described what he saw as a dreadful sight. As he wrote in 1876:

"No kid is safe from their attacks. Should a number be born together, the birds unite their forces and with great noise and flapping of their wings, generally manage to separate the weakest one and dispatch it. The birds are cruel in the extreme and the torture sometimes inflicted upon the defenceless animals is painful to witness."

These caracaras also liked to feed on shellfish, insects, worms, and rodents of all kinds, and would also go after chickens and small mammals. It is known that they often scavenged on carrion of various types, too. Their nests were usually made on the sides of cliffs, and their eggs would be laid in clutches of three.

The Guadalupe caracara was once to be found all over the island of Guadalupe, which lies off the coast of California. (This island, please note, is not the island of Guadeloupe in the Antilles, although the names of both islands are spelled and also pronounced in a similar way.)

EASY PREY

But by 1889, the Guadalupe caracaras seemed to have disappeared from the island. Many of them had probably been shot by people anxious to protect their livestock from these vicious birds. By all accounts, the Guadalupe caracaras were themselves easy prey for the islanders because they were fearless and carried on chasing their quarry, in spite of the sound of guns and the shooting of their fellow-birds. In fact, when faced with danger, a fearless caracara is said by one observer to have *"raised his crest and with an air of defiance calmly awaited death."*

But the inhabitants finally deserted the island, so no one knows for sure why the Guadalupe caracara eventually died out altogether.

Moa

Curiously, the remains of the moa, a large flightless bird, have only been found on the islands that make up New Zealand.

FACTFILE

Unable to fly, moas must have nested on the ground and are known to have laid eggs four times the size of a chicken's.

Some moas were as big as today's fully-grown giraffes – that's up to 12ft (3.6m) in height. So they were much taller than a human being. Their legs were long and slender; while their necks were very snake-like, and their pointed beaks not straight but downward-curving.

But although they certainly looked like fast-running ostriches, moas were probably not nearly so athletic and may even have been rather slow on their feet sometimes.

COLOURFUL CREATURES

According to descriptions passed down by the Maori people, natives of New Zealand, a number of moas had brightly coloured necks and a crest or comb on their heads. Indeed, some remains show areas in the skull where a crest may have grown. So it could perhaps be that these crests were only features of certain types of moa, or of male moas, or of females.

Moas were mainly herbivores, also eating occasional insects. Their strong claws must have been ideal for digging up nutritional roots; and their sharp beaks were useful for snapping at grubs or snipping off leaves and fresh plant shoots.

LIVED: probably prior to 1600
SIZE: up to 12ft (3.6m)
DISCOVERED: in New Zealand
DIET: roots, shoots, grubs, insects

OTHER DATA: scientific name, *Dinornis torosus; say:* MOH-AH; flightless; bright plumage; large eggs; much like an ostrich; long, slim legs; status: extinct

Painted vulture

Some people claim that the painted vulture never actually existed. But a naturalist who visited Florida described it in some detail before it became extinct around 1800.

Centuries ago, members of certain Native American tribes would decorate their peace pipes with exquisite wing quills taken from what was said to be the king of birds.

These feathers were from the rarely sighted painted vulture, first described to Europeans in the 18th century by the American naturalist William Bartram.

He came across the vultures in the state of Florida while travelling along the St John's River, and records this discovery in a volume of his *Travels*.

Apparently it was smaller than the more common darker vulture and not so skilled in flight. He described it as:

"*A beautiful bird, near the size of a turkey buzzard, the plumage is generally white or cream colour, except the quill feathers of the wings, which are a dark brown.*"

Bartram described the tail as large and white, tipped with dark brown or black, and the eyes had a gold iris around jet black pupils.

This type of vulture also had bare, wrinkled neck skin which was yellow and coral in colour. The bill was long and straight, but then suddenly became hooked towards the end.

SPLASHES OF COLOUR

In 1936, Bartram's original notes were found, and these seem to confirm what had been published in his *Travels*. He had also jotted down that the head of this vulture was reddish in parts and that the stomach would hang loosely on this bird just as if it was a bag or pouch. When this crop was full, however, it would protrude very prominently. Its bill was yellow, and the legs were white, he said. Indeed, it was these splashes of red, yellow, coral and purple that led to him describing the bird as "painted."

Bartram also noted that when land was set on fire in Florida for agricultural purposes, the painted vultures would gather together on the wing and descend to the burned ground. Here they would feed on a whole variety of dead creatures that had been killed off by the flames and all the smoke from the blaze.

Nevertheless, according to some schools of thought, this vulture remains merely a legendary bird.

FACTFILE

Some scientists think very severe frosts may have played a part in killing off this rare species of heat-loving vulture.

LIVED: prior to 1800

SIZE: 25.5ins (65cm)

DISCOVERED: in North America

DIET: frogs, lizards, reptiles

OTHER DATA: scientific name, *Sarcorhamphus sacra*; *say*: SARK-OR-AM-FOOS SAK-RAH; cream plumage; long bill; bald neck; red crown to head; status: extinct

Elephant bird

How terrifying the sight of a 10ft (3m) tall flightless bird must have been to the first humans to see one! It was once thought to have died out thousands of years ago, but we now know – as a result of carbon dating – that it probably survived well into the 18th century on the island of Madagascar off the coast of Africa.

It was Marco Polo, the famous 13th-century explorer, who first described this bird. It looked much like a winged dragon, he had heard, and had fiery eyes.

LIVED: prior to mid 18th century
SIZE: 10ft (3m)
DISCOVERED: in Madagascar
DIET: leaves and other vegetation

OTHER DATA: scientific name, *Aepyornis maximus*; *say*: EYE-PEE-OR-NIS <u>MAX</u>-EE-MOOS; huge eggs; possibly hunted to extinction; heavy body; status: extinct

According to Marco Polo, this bird was a gryphon, but the islanders referred to it as a rukh. It was able, he heard it claimed, to grab a fully-grown elephant with its claws and then lift it right into the air. As you can read in Marco Polo's account of his travels:

"Gryphons are large and strong enough to carry off an elephant and drop it to the ground from such a height that it will break into pieces. When the elephant has landed, the gryphon swoops down and feeds on its flesh. Those who have seen these birds say that they have a wingspan of 30 paces and feathers at least 12 paces long."

But scientists now agree that the elephant bird could not fly. What is more, elephants have never been seen on Madagascar. So Marco Polo must have been mistaken in these details about the elephant bird's behaviour, or may have been quoting from legends he had heard in that region. The rukh of the *Tales of the Arabian Nights*, for example, is said to have been strong enough to perform such amazing feats.

Many palaeontologists believe that the elephant bird (known by the scientific name of *Aepyornis*) and other giant feathered but flightless creatures sprang from a common ancestor.

Millions of years ago, when the world largely comprised one great landmass, there may have been just one species of giant bird. But when the continents formed, it is possible that groups of them became isolated and then evolved independently into different species such as New Zealand's moa (now also extinct) or South Africa's ostrich.

No one has ever found the complete skeleton of an elephant bird, but many of its fossilized bones survive intact, and fossilized eggs have been unearthed, too. These eggs are far larger than a fully-grown man's head – more than 2ft (0.6m) long – and some date from as far back as Pleistocene times. Indeed, a single egg would probably have made an omelette large enough to feed 25 people or more!

FACTFILE

As far as scientists are aware, *Aepyornis*, also known as the elephant bird, was the largest bird ever to have inhabited Planet Earth.

THE BITTER END

How, though, did this flightless bird finally become extinct? No one knows for certain. However, even though none of its remains have been discovered at ancient human settlement sites, distinct marks found on some of the fossilized bones seem to have been made by tools. So perhaps these marks are proof that their flesh was removed and eaten by humans. Their eggs may have been stolen and enjoyed, too.

Heath hen

The heath hen, extinct today, was once so common on the dinner tables of North America that it would sometimes be served as often as three times each week.

Millions of heath hens once roamed wild over the plains of New England; but as soon as early settlers started shooting them for food, numbers decreased rapidly. These birds were also vulnerable to attack by cats and dogs, and even large rats, as they were ground-dwellers. Goshawks and other birds of prey were attracted to them, too, and would swoop down to grab their prey. Heath hens had slow reactions and almost always failed to escape the clutches of these very greedy predators.

Human beings were both directly and indirectly culprits, too. Immigrants had brought with them new types of poultry, and together with these came severe epidemics of disease that killed off heath hens in huge numbers.

The natural habitat of heath hens, meanwhile, was gradually being destroyed by conversion to arable land. In the late 18th century, an attempt had been made to pass a law to prevent too many being killed but few people cared.

FACTFILE

Some heath hens caught a severe poultry disease called Blackhead, and their numbers dwindled.

When heath hens finally vanished from the mainland, about 200 remained on the island of Martha's Vineyard in the state of Massachusetts. Here, though, they also started to decline until a reservation was set up. Yet, by1896, the number of heath hens on the island had decreased to under 100; and within a few years, this figure halved.

ATTEMPTS AT CONSERVATION

A few determined individuals then decided to set up a 1,600-acre (648-hectare) conservation area on Martha's Vineyard for the protection of the remaining heath hens; as a result, in just seven years, the population of these birds increased as much as fiftyfold. Special wardens were even employed to see to their welfare and to check for poachers.

At that time, however, there were considerable changes in climate. Bush fires rapidly destroyed the heath hens' breeding grounds, and they suffered badly in a very harsh winter. Records show that by 1927 there were only 13 surviving heath hens on the island; and by 1928, there were just two. The very last one narrowly escaped being run over by vehicles failing to notice it strutting across the road, until it finally died in 1932.

LIVED: prior to 1932
SIZE: chicken-sized
DISCOVERED: in North America
DIET: mostly seeds and whatever available

OTHER DATA: scientific name, *Tympanuchus cupido cupido*; *say*: TIM-PAN-OO-KOOS COOP-EE-DOH COOP-EE-DOH; bush fires destroyed its breeding grounds; status: extinct

World of the heath hen

- The heath hen was related to a similar bird, the prairie chicken, which still survives in North America.
- The heath hen had only a very small brain and was not intelligent.
- The flesh of the heath hen was very tasty, yet people got bored with it because it was so readily available and served so often for dinner.

Other lost animals

To find out about an extinct goat that was only the size of a small dog, an antelope that seemed to have a blue coat, and much more, dig into this section of the book.

Creatures as varied as the dwarf elephant of Sicily, the quagga of southern Africa with its partially striped coat, the Tasmanian tiger, the Falklands fox, a horned turtle, and the giant beaver of North America all feature on the pages that follow.

Most if not all of them will be unfamiliar to you because, like all the animals and birds described in this volume so far, they are no longer to be seen in the wild, even though a few have relatives alive now that resemble them in several respects.

HORROR STORIES

The reasons why each of them have finally disappeared make for fascinating reading, but for nature-lovers they are, in some instances, tantamount to horror stories. Why have we treated so many of our fellow creatures with such disdain?

Need we, for instance, have lost the West Indian monk seal if people had not been so greedy for its skin and its meat? Similarly, could Dawson's caribou, a delightful type of deer, have been saved from extinction with just a little more forethought? Could we also perhaps have taken a leaf out of Noah's book and saved some of these beasts when various forms of disaster struck them?

One of the greatest thing about life on Earth has always been the variety of the species. It is hard to credit that there could be so many different creatures, and we can only wonder at it all. Yet conservationists keeping track of the declining numbers of some animals reckon that a frighteningly large number of them continue to disappear altogether from the planet each year. Clearly we still have to learn some vital lessons from extinctions of the past.

Tasmanian tiger

What an extraordinary creature the thylacine was! Its body covering was like a tiger; yet, in one respect, it resembled a kangaroo.

Imagine the scene as a scrawny animal paced its cage, and visitors stared. Perhaps, deep down, it knew that it was the last of its kind. The thylacine (or Tasmanian tiger) certainly looked rather like a tiger with all its stripes; but it was a meat-eating marsupial – the largest of its kind – with a generous pouch that could hold up to four young.

PAID TO KILL

The world's last thylacine died behind bars, in Beaumaris Zoo, Hobart, Tasmania, in September 1936. Once common in Australia and New Guinea, these creatures found it hard to survive the threat of dingos (wild dogs). Those that were native to the island of Tasmania, meanwhile, threatened sheep, with the result that the Tasmanian government offered the incentive of generous payment to farmers for every thylacine killed.

Since 1936, some claim to have spotted the occasional Tasmanian tiger in the wild, but such reports have not been confirmed. This extraordinary creature is, it seems, yet another of the world's sadly lamented vanished species.

LIVED: prior to 1936
SIZE: up to 6ft (1.8m) long
HABITAT: Australia, Tasmania, and New Guinea
DIET: carnivorous

OTHER DATA: backward opening pouch of an opossum, therefore a marsupial; scientific name, *Thylacinus cynocephalus; say:* THEYE-lASS-EE-NOOS SEYE-NOH-SEF-AL-OOS; status: extinct

Quagga

Since 1878, the world has mourned the very last quagga to be killed in the wild. London Zoo once had two specimens, but they did not breed in captivity.

It's a sad thought that no one will ever hear the cry of a live quagga again. Unfortunately, it became extinct before humans had learned how to record sound. But scientists do have a rather good indication of what this was like because the quagga's name is known to be derived from its call.

Early Dutch settlers, reaching the part of Africa where the quagga lived, first called this animal the *quahah* (KWAH-HAH), in imitation of the noise it made when communicating. Only later did the name get changed to "quagga."

Try saying *quahah* out loud. Now repeat this noise several times very deep in your throat, and you will probably sound just like the quagga did!

If you had looked at the quagga from the front, you might have mistaken it for a zebra. After all, it was the right shape and seemed striped. But if you saw it from the rear, you might well have changed your mind. A little way down its back, the stripes suddenly faded to an all-over colour.

FACTFILE

Settlers in southern Africa hunted quaggas for their hide, which was strong but light, and thought to be ideal for bags.

LIVED: prior to 1878
SIZE: similar to a zebra in height
HABITAT: southern Africa
DIET: vegetation
OTHER DATA: hunted for its hide; attacked by jackals; kept to guard livestock; scientific name, *Equus burchelli quagga; say:* EK-WOOS BER-CHEL-EE KWAG-AH; status: extinct

Early settlers in southern Africa needed to protect their livestock as there was always risk of attack by all sorts of wildlife – jackals, for instance on the prowl for sheep, cows or chickens.

Many farmers actually succeeded in training quaggas to act much like guard dogs. Not only would the quaggas, with their excellent hearing, call loudly at the approach of any carnivorous beast or human poacher, they would also attack quite violently. It did not save them.

Blue antelope

The very last blue antelope is thought to have been seen in South Africa around the year 1800. It is said to have had a mysterious blue-grey tint to its skin at certain times, hence its name; and according to some authorities it was probably a close relative of the sable and the roan antelopes.

Take a good look at the illustration shown here. Does it remind you of another creature altogether? Yes, some authorities say the blue antelope had a small beard and horns, just like a goat's.

Hunted primarily for its skin, the blue antelope was first described in 1731. At that time, it was already rare; and just 69 years later it became extinct. Also known as the blue buck or *blaauwbok* (its name in the Afrikaans language), the blue antelope remains mysterious to scientists in many respects.

LIVED: prior to 1800
SIZE: similar to a deer
HABITAT: grasslands of Cape Colony, South Africa
DIET: vegetation

OTHER DATA: its meat was fed to dogs; also called a blue buck or *blaauwbok*; scientific name, *Hippotragus leucophaeus; say:* HIP-OT-RAHG-OOS LOOK-OFF-EYE-OOS; status: extinct

SHADES OF CONFUSION

Reptiles and birds sometimes have blue, or perhaps just a tinge of this shade, as part of their colouring; but it is hard to think of a mammal that could be described as being blue.

One theory, however, is that the bluish appearance of this antelope only occurred in older members of this species. When their coats gradually thinned with age, the bluish skin underneath, it is thought, started to show through.

The blue antelope is said to have been smaller than the roan and sable antelopes, and it was lighter in weight, too.

It liked to inhabit fairly damp regions, and this has suggested to scientists that it must have needed to drink at regular intervals. In this, like the oran and sable antelopes, it would have differed from certain other types which are able to extract water from the plants they eat.

FACTFILE

The blue antelope was hunted extensively; but its meat was not enjoyed by humans. Instead, it was used for dog food.

At one time, because accounts of the appearance of the rare blue antelope were so confusing in their details, some scientists suggested that it perhaps had not been a separate species from the roan or the sable at all. It seemed that reports describing its physiology might have been inaccurate.

However, remains show that there were indeed several differences, in addition to colour and size, between these antelopes. The premolar teeth of the blue antelope, for instance, were definitely longer than those of the sable. Its horns, meanwhile, were more like the sable's than the roan's.

Long premolars generally belong to creatures that browse a lot on grass, so the blue antelope probably fed mostly in this way, only occasionally munching broad leaves.

Today, alas, both the roan and the sable antelopes have become endangered. Conservationists hope to act in time to save them.

Syrian wild ass

There are carvings that show the Syrian wild ass being caught as early as 650 B.C. It was mainly the young of this species that were killed for their meat, until the very last of these animals died in the wild in 1927.

Also known as the Syrian onager, the animal, shown *right*, inhabited a far wider area than just Syria, in spite of its name, and once lived in Palestine, Arabia, Iraq, and beyond. Here, for at least two thousand years, it was hunted by Man but often managed to survive due to the speed with which it could escape human predators. It was also very well adapted to a desert environment.

The Syrian wild ass was very small, only measuring about 3ft (90cm) in height at the most, and was light in weight, too. Because of this, it was never used for carrying goods.

Sometimes the Bedouins would raise the young of the onagers in their tents, in preparation for slaughter and a subsequent feast.

Once, guns and vehicles reached the Middle East, however, the Syrian onager became more readily run down, such activity reaching a peak during the First World War (1914-1918).

FACTFILE

In the 16th century, some Middle Eastern nomads succeeded in taming a few female Syrian wild asses and would milk them.

SURVIVING RELATIVES

Today, there are other types of ass that are not yet extinct but that have become severely endangered. One is the Asiatic wild ass from Mongolia. The Persian wild ass, meanwhile, is now only found in a small area of Iran and part of Russia, where a reserve has been established especially for it. The Indian and the Tibetan species of wild ass, known as the kiang, are also in decline.

There are two types of African wild asses. The Somalian is found in Ethiopia, Eritrea, and north Somalia, where constant wars and political upheavals have made sightings of these creatures increasingly scarce.

The Nubian ass, which is the other African species, is also rare and found in a small part of Sudan and Eritrea. However, no one is entirely sure whether these surviving creatures are truly the original wild species. They might be domesticated asses that have somehow managed to escape, or that have been released into remote regions.

From time to time, there are also reports that the onager has been sighted in Saudi Arabia or Oman; but the chances are that these claims are inaccurate and that the asses seen were also domesticated creatures, turned wild.

LIVED: prior to 1927
SIZE: 3ft (0.9m) tall
HABITAT: desert regions in Syria, Palestine, Arabia etc
DIET: vegetation

OTHER DATA: slaughtered for meat; not a pack animal; also called an onager; scientific name, *Equus hemionus hemippus; say:* EK-WOOS HEM-EE-OHN-OOS HEM-IP-OOS; status: extinct

Grey's wallaby

In a scene from the beginning of the 20th century in the Australian outback, shown *right*, a Grey's wallaby stands rooted to the spot as five fierce dogs confront it. Read on to find out what happens next.

The dogs start to bark furiously; yet for some inexplicable reason the wallaby does not run off in haste but stares hard at the dogs, almost as if issuing a challenge to them to give chase. It stands perfectly still for a few seconds and then suddenly turns, bounding off at a terrific rate. It is like greased lightening.

In such a situation, however, every second counts, and this wallaby has miscalculated. It has rapid reactions but the dogs are also quick off the mark and have been trained in predatory skills. They soon catch up and bring the unfortunate marsupial down.

The dogs' owners will now lay claim to the Grey's wallaby's skin, and its meat will be enjoyed, too.

The Toolache wallaby, another name for the Grey's wallaby, had a lengthy tail which was almost as long as the rest of its body.

Reputedly very fast on its feet, it would often take the risk of waiting until the very last moment before bounding away at a tremendous rate to avoid being caught, as described *above*.

RISKY TACTICS

This was a characteristic that must undoubtedly have contributed to the demise of Grey's wallabies. Time was clearly of the essence even for such a speedy creature as Grey's wallaby, if it was to remain alive.

Another factor that almost certainly had a part to play in the road to extinction was their preference for open ground as a habitat. This meant there was rarely anywhere for them to hide when threatened.

The general consensus of opinion is that the Grey's wallaby is extinct. But another type of wallaby – the white-throated species – was also thought to have died out in 1932 in a region of New South Wales, Australia, where it was once plentiful. Destruction of its natural habitat was to blame. Then, in 1966, entirely unexpectedly, this wallaby suddenly reappeared across the water near Auckland, New Zealand.

Experts were confounded at first. How could Grey's wallaby possibly have reached these shores? Then it was discovered that a number of these wallabies had in fact been introduced to the wild by settlers in New Zealand at the end of the 19th century. Amazingly, too, in 1972, a small number were rediscovered in a remote area back in Australia. Perhaps we are sometimes a little too rash in declaring a species extinct, it could be said.

LIVED: prior to 1940
SIZE: up to 3ft (0.9m) long
HABITAT: Australian outback
DIET: vegetation

OTHER DATA: marsupial; hunted for its skin; speedy on its feet; also known as the Toolache wallaby; scientific name, *Macropus greyi*; *say:* MAK-ROH-POOS GRAY-EE; status: extinct

World of the Grey's wallaby

- There was once so much demand for the pelts of Grey's wallabies that anyone trying to make money was tempted into hunting them for financial reward.
- Their skins would be put to a whole variety of uses, including the manufacture of items for the fashion industry. There was a market for their meat, too.

FACTFILE

The last of the Grey's wallabies died in Adelaide Zoo, Australia. These creatures did not breed well in captivity.

Zaglossus hacketti

Spiny ant-eaters, like the one in the illustration *below*, will use their long, slender snouts to sniff out a meal of the insects that form part of their name and also comprise these creatures' staple diet. Scenes like this take place in remote parts of Australia and southern New Guinea today. But long ago there was also a *giant* spiny ant-eater.

LIVED: as far back as Cretaceous times
SIZE: 3.3ft (1m) long
HABITAT: Australia
DIET: ants and termites mainly

OTHER DATA: also known as a spiny ant-eater or echidna; a monotreme; long tongue; scientific name, *Zaglossus hacketti*
say: ZAG-LOSS-OOS HAK-ET-EE; status: extinct

Known to scientists as *Zaglossus hacketti*, this ancient spiny ant-eater (also called an echidna) existed on Earth at a time when dinosaurs were still roaming the planet. Like today's echidna, which it resembled closely, it would have had a very long and sticky tongue which was ideal for lapping up the ants and termites on which it lived.

The coat of this ancient ant-eater, as you can see, consisted of numerous coarse hairs and spines which provided at least some degree of protection against predators and might also have camouflaged the creature well among the brush.

The life span of a spiny ant-eater of today is long for an animal of this type, even in captivity, so we can hazard a guess that this applied to *Zaglossus hacketti*, too. It may, however, have fallen victim to large predators with which it coexisted long ago.

Zagolossus hacketti was also what scientists call a *monotreme*. (This word means "single hole," and refers to the urinary, digestive and reproductive organs that have a single opening.)

The echidna of today is also a monotreme, as is the duck-billed platypus. But *Zaglossus hacketti*, as we can tell from remains, was about twice the size of the modern 20-inch (50-cm) long echidna.

EGG-LAYING MAMMALS

Monotremes are most unusual creatures because, although strictly speaking they are mammals as they provide milk for their young, they lay eggs rather than giving birth to live offspring.

Today's echnidna will lay a single egg that hatches into a pouch, and we can only assume that Zaglossus did the same. Like today's baby monotremes, it may also have obtained nourishment by licking at the milk that its mother excreted into the pouch.

For some time, a scientist called Peggy Rismiller has been studying the wildlife on Kangaroo Island, near Adelaide, southern Australia. She has a particular interest in the echidna population there, and as a result of her research may well come up with a few more clues as to the life of their ancient, far larger relative.

World of Zaglossus hacketti

- Today's echidna is a shy creature and uses its strong claws to dig into the ground and hide if danger threatens. Although *Zaglossus hacketti* was larger, it may well have gone underground, too, until the coast was clear.

Potoroo

There have been several species of potoroo, some of which are extinct in certain parts of Australasia. The broad-faced potoroo, shown in the illustration, for instance, finally disappeared from Western Australia in the early 20th century.

John Gilbert, the great 19th-century British explorer and naturalist, collected a specimen of the broad-faced potoroo when it was already rare in its natural Australian habitat. There were no potoroos, also known as rat-kangaroos because of their close resemblance to rodents, to be found in Europe. Eventually it was to become extinct in about 1908, partly due to the effect of severe bush fires which literally burned them out of existence in the wild.

Gilbert also had a species of potoroos named after him; but the so-called Gilbert's potoroo had vanished completely from Western Australia a few years earlier, in 1900. This potoroo had a black tail and a characteristic black stripe running centrally down its face; and from Gilbert's account we know it lived in thickets by running streams.

Its natural habitat was gradually destroyed, however; and it was preyed upon by foxes. It seems, too, that the local population would kill it by first driving it into the open and frightening the creature with loud noises. Then they would attack it with a spear. Gilbert had discovered this species of potoroo at a place called King George's Sound in Western Australia in 1840.

LIVED: some types survive, others extinct prior to 1908
SIZE: up to 2ft (60cm) long
HABITAT: marshes and water courses in Australasia
DIET: vegetation and scraps

OTHER DATA: marsupials; preyed on by cats, foxes, and humans; long tails; scientific name for Gilbert's potoroo, *Potorous gilberti*; *say:* POT-OR-OOS GIL-BERT-EE; status: extinct

ANCIENT TINY KANGAROO

There is also another type of this creature, known as the long-nosed potoroo. It is in fact Australia's oldest real kangaroo in spite of its diminutive size – it measures just 8ins (20cm) – and has even been described as a virtual living fossil.

This is because some scientists think it probably looks just the same as it did many million years ago. But it has not been seen in Western or South Australia for very many years, and must be extinct in those regions. Today it is known to exist for sure only on the island of Tasmania.

World of the potoroo

- Cats were introduced to the area where the St. Francis Island potoroo was found towards the end of the 19th century. This was done to keep the potoroo population down and to prevent them from destroying home-grown vegetables. However, so many of this species were killed by feline predators that the native potoroo became extinct on St. Francis Island, off the coast of South Australia, by 1900.

Extinct bandicoots

Although some bandicoots survive in Australia, several species of this marsupial became extinct during the 20th century. The fur trade has been partly to blame. It is certainly sad that such cute creatures as these endearing bandicoots are gone forever.

Discovered in 1840 but extinct just 100 years later, prior to which it had lived in the states of Victoria and New South Wales, the Eastern barred bandicoot, shown here, still lives on the island of Tasmania. Across its back are the characteristic stripes or bars that feature in the animal's name.

FACTFILE

Many so-called greater-rabbit bandicoots are thought to have been killed when they were caught in traps actually set for rabbits.

Some of the difficulties surviving eastern barred bandicoots face in Tasmania, and that their extinct cousins from the mainland must have come up against, too, include predation by foxes, cats and dogs. Cats may also carry a disease known as *Toxoplasmosis* which can be caught by the barred bandicoot. It is usually fatal.

The clearing of grasslands for agricultural use is leading to a decline in their numbers, too, even though they reproduce prolifically. A single female may give birth to 16 offspring in one year.

LIVED: some extinct by 1940
SIZE: smaller than rabbits
HABITAT: grasslands in Australia
DIET: mostly vegetation

OTHER DATA: natural habitat destroyed; scientific name for Eastern barred bandicoot, *Perameles fasciata; say:* PER-AM-EEL-EES FAS-KEE-AH-TAH; status: extinct

World of the barred bandicoot

- One of the prime reasons why some types of barred bandicoots became extinct is that they did not burrow at all. They therefore became victims of bush fires, and had nowhere to hide when their cover of vegetation was lost.

The Western barred bandicoot was last recorded on the mainland in 1922. It had an additional single stripe at its sides, just by its thighs, and also a backward-opening pouch. Today, this bandicoot is only found on the islands of Bernier and Dorre, where they were rediscovered in 1983. Here, they live mainly in sandhills behind beaches or in dunes covered by sparse vegetation. They are principally nocturnal creatures, and feed mainly on vegetation and insects.

The pig-footed bandicoot, meanwhile, was given its name because its front limbs closely resemble pigs' trotters. At the same time, its long, slim legs give it the look of a tiny deer. It was also from South Australia, and seems last to have been spotted in 1925.

Bilbies, otherwise known as greater rabbit-bandicoots, were once to be found in South Australia, too, but have now entirely disappeared from this state. Early settlers found them delightful and did their best to protect them. But, in time, people lost interest and economics took over so that they were killed off.

Dawson's caribou

This species of deer once lived in swampy regions on the Queen Charlotte Islands, off British Columbia, Canada, and probably became extinct there in 1908.

Found only on Graham Island, which is part of the Queen Charlotte Islands group, the Dawson's caribou probably died out at least partly due to hunting but also because of changes to its natural habitat. The Haida Native Americans, who had lived on the Queen Charlotte Islands for thousand of years before immigrants arrived, were not even aware of its existence until this was pointed out to them. These Indian peoples rarely ventured to central regions.

FACTFILE

Dawson's caribou is named after G. M. Dawson, the man who, in 1878, first wrote about them.

LIVED: prior to 1908
SIZE: approximately 5ft (1.5m) to shoulder
HABITAT: Queen Charlotte Islands, Canada
DIET: vegetation

OTHER DATA: hunted for pelts and meat; lived in swampy regions; scientific name, *Rangifer tarandus dawsoni*; *say*: RAN-JEE-FAIR TAH-RAN-DOOS DOOR-SOHN-EE; status: extinct

World of the Dawson's caribou

- In 1908, after only occasional sightings, two male and a single female Dawson's caribou with just one baby were seen. The mature creatures were killed, but the baby got away. Luckily, the young were always too shy to approach humans.

However, this changed completely when the Haida began to assist immigrant collectors in hunting the caribou for their pelts. These caribou were inquisitive by nature and could be brought within close range if a hunter tried to attract them.

Dawson's caribou had long, thick winter coats which provided the animals with warmth during the coldest months but which the hunters sought.

Certain types of caribou are known to be excellent swimmers, and so it may have been that the Dawson's caribou originally came to Graham Island across the water from the mainland, although no one can be entirely certain.

THREATENED COUSINS

There are other types of caribou that, right now, are reported to be declining in number, among them the Peary caribou, found only on islands lying in the Arctic archipelago.

Some species of caribou, meanwhile, remain a source of food for people living in remote northern settlements. However they do not seem to be endangered, fortunately, and tourists flock to see them.

Nevertheless, conservationists will need to keep a careful eye on them if the fate of Dawson's caribou is to be avoided.

American bison

Millions of bison once roamed the North American prairies; but hunting brought them to near extinction in the late 19th century. The story has an unexpected ending, however.

About 200,000 years ago, ancestors of the bison crossed a land bridge from Asia and arrived in North America. At that time, these creatures were much bigger than the bison known today, and may have weighed more than twice as much. They thrived in this new environment but decreased in size over many thousands of years.

Once Native Americans had made contact with bison, they began to hunt them extensively, not only for their meat but also for their hide, from which they made tepees, blankets, shoes, and clothing, while the bones, horns, and hooves were used for tools.

World of the American bison

- The herds of bison that lived on the great plains before this population was decimated are estimated to have numbered about 60 million. But by 1860, there were only about 850 of these animals left.

KILLING SPREES

It was with the arrival of more and more Europeans in the so-called New World, however, that wholesale slaughter of the bison began to take place. Buffalo Bill, whose real name was William Frederick Cody, is even said to have killed over 4,250 in just 18 months so that there would be enough meat to feed all the men who were building railroads across the continent.

Many bison were also lost because they would stampede into swampy ground in a desperate attempt to escape their human predators. It is even said that the corrupt government had the bison killed in huge numbers because they wanted to destroy the economy of the Native Americans and so get the Plains Indians under their control.

It also became something of a status symbol among new immigrants to eat bison flesh or to have something made from its skin. Sad to tell, you could buy a ticket for a special seat on a train to shoot at the bison through the window. The aim was wholesale slaughter.

Just in time, however, the American bison was rescued from a severely endangered status and given legal protection.

LIVED: from 200,000 years ago until the present
SIZE: today's bison smaller than previously
HABITAT: first Asia, then North American prairies
DIET: vegetation

OTHER DATA: once found in huge numbers but hunted for their hides and meat; bones, horns, and hooves also made into tools; status: rescued from complete extinction

THE BISON ARE BACK!

Today, thanks to successful conservation programmes, there are thought to be about 350,000 bison on ranches and in national and state parks in North America. This number is, of course, only a fraction of the original population but, nevertheless, this magnificent creature has been saved.

Even if you are unable to get to see bison in the wild – in Yellowstone National Park, Wyoming, for example – it could be you will see some in your local zoo.

Giant beaver

The modern beavers of North America, which are now endangered, are very much like their distant cousins, the giant beavers that died out about 10,000 year ago. But there was one main difference – size. Fossils in the Field Museum, Chicago, show they were about 9ft (2.7m) long.

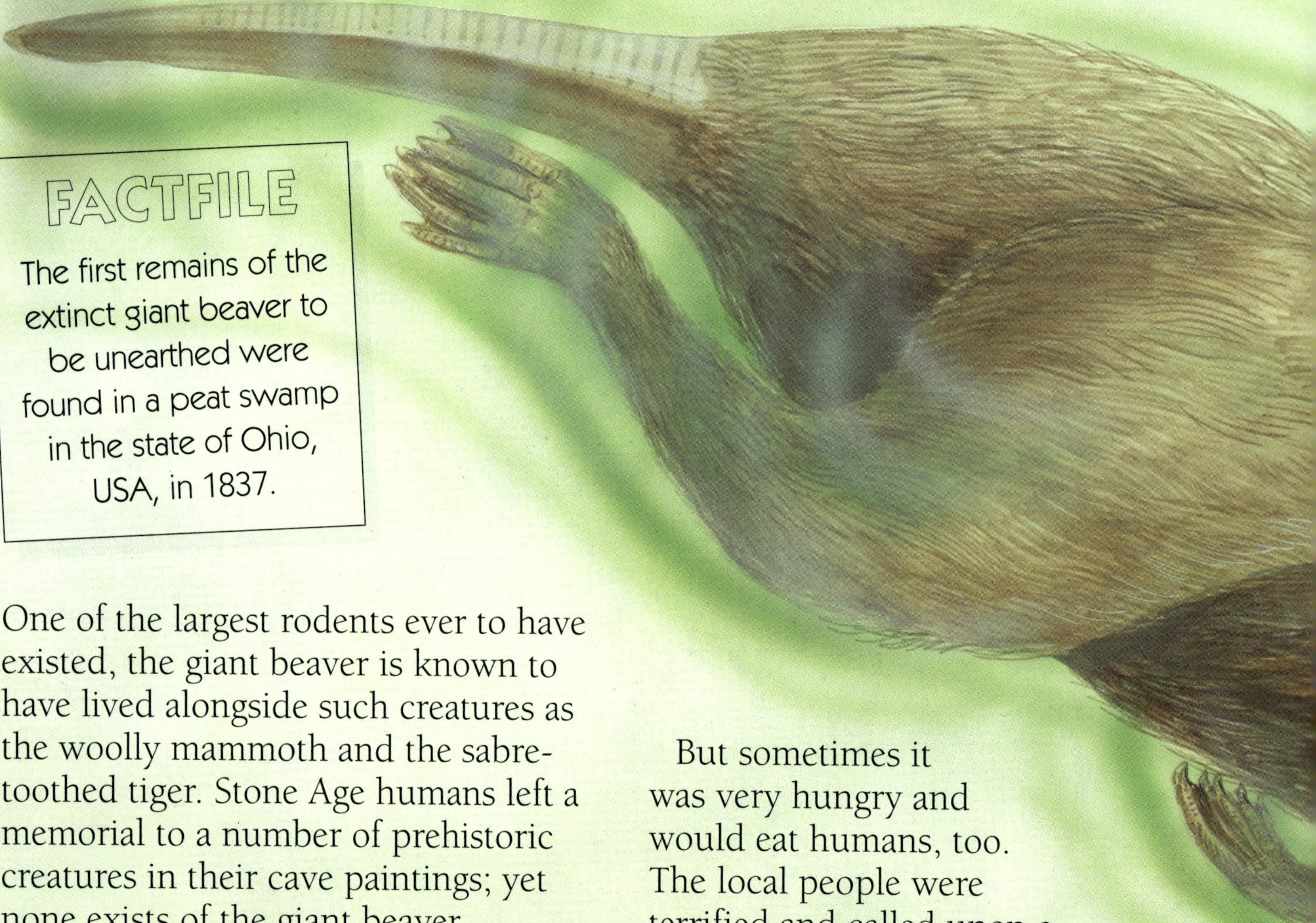

FACTFILE

The first remains of the extinct giant beaver to be unearthed were found in a peat swamp in the state of Ohio, USA, in 1837.

One of the largest rodents ever to have existed, the giant beaver is known to have lived alongside such creatures as the woolly mammoth and the sabre-toothed tiger. Stone Age humans left a memorial to a number of prehistoric creatures in their cave paintings; yet none exists of the giant beaver, Scientists have therefore had to rely entirely upon fossils for a picture of what this massive rodent must have looked like. Native American legends also provide clues as to how awesome it must have been.

According to an intriguing Indian belief, the giant beaver preyed regularly on the fish living in Long River.

But sometimes it was very hungry and would eat humans, too. The local people were terrified and called upon a kindly spirit giant, Hobomuck, to kill the beaver. Hobomuck agreed to oblige and chased the beaver into the depths of a great lake. Then he threw rocks at it and managed to sink the creature before hitting it with a great club. To this day, it is said that its head became the sandstone cliff of Mt. Sugar Loaf, and its body, the northern range.

LIVED: more than 10,000 years ago
SIZE: 9ft (2.7 m) long
HABITAT: North America
DIET: mainly fish

OTHER DATA: subject of many Native American legends; lived in rivers and lakes; distant relative of modern beavers; first remains found in a peat swamp; status: extinct

World of the giant beaver

- There is no evidence that humans hunted the giant beavers which finally died out about 10,000 years ago.
- A baby beaver is called a kit, and a beaver's home is called a lodge.

The hollow that lies between, meanwhile, is where Hobomuck's cudgel hit the giant beaver's neck.

This legend of the Pocumtuck tribe may possibly have some basis in truth because we do know from fossil evidence that a very large lake did indeed cover the Connecticut River Valley, and that a species of giant beaver once lived in that lake.

THREATENED EXTINCTION TODAY

Modern beavers have also become severely endangered in some areas, principally because of trade in their pelts. In the 19th century, these were used for top hats and collars. More recently, beavers were also killed so that the fashion trade could turn their skins into fur coats. However, some people have killed them because they considered them pests. Humans can catch a disease from beavers if they happen to drink water that has been contaminated by these otherwise enchanting rodents.

In Europe, the modern beaver is almost extinct. Now, however, the trapping of beavers is limited to certain seasons of the year by law in some regions to help prevent any further risk of extinction of this species.

Dwarf elephant

We usually think of elephants as large and bulky. So would you believe that there were once dwarf elephants that stood only about 3ft (1m) tall as adults? Fossils of these elephants have mainly been found on the island of Sicily. Just imagine how tiny their babies must have been!

Why is it that such a tiny species of elephant evolved while, elsewhere, elephants remained the large creatures we know today? There is a theory that explains why diminutive species, such as the dwarf elephants, are sometimes found only in an island environment. Once these elephants had swum to Sicily and established a new habitat, they would have found limited resources in a reduced habitat.

FACTFILE

Some dwarf elephant remains, found in caves, date back to the Pleistocene era. But they may not have died out until Roman times.

There were also probably no giant predators. So the immigrant elephants did not need to maintain their original body size and became smaller in time.

GIANT MYTHS?

It may even be that dwarf elephants lie behind accounts of the giant rokh, or elephant bird, that are sometimes thought to be grossly exaggerated or even simply myths. Perhaps, after all, this huge bird may have been able to lift elephants – dwarf elephants, that is, or maybe only baby dwarf elephants – right into the air, then dropping them to their death. (You can read more about the elephant bird on pages 134-135.)

LIVED: in Pleistocene and more recent times
SIZE: 3ft (less than 1m tall)
HABITAT: Sicily
DIET: vegetation

OTHER DATA: just like miniature versions of those elephants we know today; scientific name, *Elephas falconeri; say,* EL-EF-AS FAL-KOH-NER-EE; status: extinct

Some experts believe that the famous Greek myth about the one-eyed cyclops, Polyphemus, in Homer's epic, *The Odyssey*, originated due to the early discovery of fossilized remains of dwarf elephants. (This was long before it was realised what these actually were, of course.) The fossilized skulls in particular could well have led to a major misunderstanding. There was a large central hole, which might have been thought to be the site of the huge central eye that the cyclops was said to have had.

Rat-rabbit

The chances are that you find the thought of eating rats absolutely disgusting. But the so-called rat-rabbit, also known as the pika, was actually a *lagomorph*, as rabbits and hares are, and not a rodent.

If you look in a traditional French cookbook, you will find recipes for dishes that include rabbit or hare meat. So it is hardly surprising that, way back in time, inhabitants of the French island of Corsica caught the local pikas, or rat-rabbits, for their flesh.

To judge from its fossilized remains, the Corsican pika must have been very common during the time of the last Ice Age, and early humans alive then probably ate a lot of its meat when other larger animals were scarce.

Corsica became inhabited about 9,000 years ago; and the pika, also once found on the island of Sardinia, is thought to have become extinct in about A.D.500. From this we can gather that it is also likely to have featured on the menu during Roman times. No one can be sure today, but it probably tasted just like rabbit or hare when prepared on a primitive spit-roast.

FACTFILE

Several distinct types of pika exist today and are found in habitats as varied as deserts, rocky terrain, open plains, and steppes.

LIVED: prior to 500 AD
SIZE: up to 12ins (30cm) long
HABITAT: rocky terrain on Corsica
DIET: grass and other vegetation

OTHER DATA: probably became extinct due to overhunting; scientific name, *Prolagus corsicanus; say:* PROH-LARG-OOS KOR-SIK-AH-NOOS; status: extinct

World of the rat-rabbit

- If the extinct rat-rabbits of Corsica were like the pikas that exist today, they produced green droppings during the day but black droppings at night. They may also have eaten some of their waste as a way of retaining nutrients.

According to one theory, the pikas, or rat-rabbits, originally came from the mainland of what is now Europe at a time when there was a land bridge, and remained on the islands of Sardinia and Corsica when the sea levels rose and there was no return route.

Some of these extinct rat-rabbits are thought to have grown to more than 12ins (30cm) in length. They probably moved very quickly and must have been difficult to catch, though their human predators probably used spears and also set traps of various kinds.

But why did the rat-rabbit of these times die out? It was almost certainly due to the introduction of species that, in addition to humans, became its predators.

Lost goats

Fossilized remains of cave goats have been discovered on the Mediterranean island of Majorca. They may have started to live there as long as 8 million years ago.

The cave goats of Majorca, depicted across these two pages, do not seem to have been found elsewhere, and were tiny in comparison with the type of goats with which we are familiar today. In fact, a mature cave goat was only about 20ins (50cm) tall, which is the approximate height of a 20th-century goat kid.

But, then, as also explained on page 162, it is frequently the case that animals moving to live on islands will evolve to be smaller than they were when living elsewhere if there are no dangerous predators sharing their island habitat.

FACTFILE

Cave goat horns, according to some authorities, may have been taken by humans to use as amulets or lucky charms.

INSIDE INFORMATION

Cave goat skulls have been found with filed horns, providing an almost certain indication that they would have been kept as domesticated animals by the early humans with whom they may have shared cave dwellings. There are also caves that contain their fossilized excrement; and some, too, with wall paintings that show goats closely resembling what we think they must have looked like.

But another type of goat has become extinct much more recently. In January 2000, it was reported that a protected species of goat, indigenous to a mountainous region of northern Spain, had become extinct when the last known living specimen was found dead under a fallen tree. It was a 13-year-old female, and its skull had been crushed.

DATA: the reasons why certain types of goat have sometimes become extinct from the prehistoric era right through to more recent times include hunting; environmental changes; predation by other animals; possibly disease and natural disasters; the taking of their horns for use as talismans

Having lived in a national park for some years because this Pyrenean mountain goat had gone into decline due to environmental changes and the effect of poaching, it has now disappeared. But perhaps not forever because scientists have taken a tissue sample in case they decide to clone the creature. However, this would only work if a surviving male is found. The successful cloning of another species of goat took place in America in 1999.

Falklands fox

Strangely, at one time, some Falkland Islanders who shared this creature's habitat believed it was a vampire. But this is just one of several reasons why it was hunted to its eventual extinction in 1876.

Sometimes called the Antarctic wolf or the warrah, the Falklands fox was long persecuted on the Falkland Islands, which lie off the coast of Argentina, South America. Here they were killed after being lured with a piece of meat, and then stabbed.

The group of Falkland foxes in the illustration across these two pages are seen running for their lives to escape a fire. Back in 1764, a few sailors were attacked by a Falklands fox, which they described as the only savage animal and quadruped in the region.

As a result, the men deliberately lit an extensive area of grass to try and drive the foxes from one particular part of the island where they had dropped anchor. The blaze is said to have lasted for several days, and no doubt many of the foxes did not survive the horrific inferno.

LIVED: until 1876
SIZE: like other foxes
HABITAT: Falkland Islands
DIET: carnivorous

OTHER DATA: hunted for its pelt; shot as killed sheep; also called an Antarctic fox or warrah; scientific name, *Dusicyon australis; say:* DOOS-EES-EE-ON OW-STRAH-LIS; status: extinct

The famous 19th-century British naturalist Charles Darwin saw the Falkland fox for himself when he sailed to this part of the world, and he was to predict that this animal would no doubt become extinct before too long.

In the *Zoology of the Voyage of the Beagle*, he wrote that:

"The number of these animals during the past fifty years must have been greatly reduced; already they are entirely banished from that half of East Falkland which lies east of the head of San Salvador Bay and Berkeley Sound; and it cannot, I think, be doubted that as these islands are now being colonized, before the paper is decayed on which this animal has been figured, it will be ranked among those species which have perished from the earth."

FACTFILE

A single specimen of the Falklands fox was brought to England in 1868, where it finally died at the London Zoo.

The Falklands fox was hunted for its pelts which were turned into fur coats for the international fashion trade. It was also said by some farmers on the islands to have killed a lot of sheep, and so was shot whenever possible to keep numbers down.

Most mysterious, however, was the fact that it was the only predatory beast (apart from a small mouse) on the islands. When it could not catch a sheep, it must have lived on penguin meat, eggs, and perhaps vegetation. There was certainly a shortage of food for it, and one report says that it was common for the Falklands fox to starve for part of the year.

World of the Falklands fox

- No one is sure how this creature reached the Falkland Islands, which are 180 miles (300km) from Argentina. One theory is that they arrived in Pleistocene times, when the sea level was much lower; another, that they were once tame.

W. Indian monk seal

In 1495, Christopher Columbus and his crew reported killing eight West Indian monk seals, which they described as "sea wolves." Constantly hunted for their blubber, skins and meat, they are now extinct.

This seal was in fact the first New World animal to be seen by the great explorer Christopher Columbus when he reached the islands of the Caribbean waters. It moved extremely slowly and was very trusting towards humans. This, some scientists have suggested, was one of the principal factors that finally led to its disappearance.

These seals would come ashore to breed, and here they would be particularly vulnerable. Hunters could creep right up close, appear friendly, and then suddenly slaughter them on the spot. Because of the way they lumbered along on land, the monk seals did not stand a chance. Scientists were also sometimes to blame for killing seals simply because they wanted them for their museum collections.

The West Indian monk seals were brown-grey in colour but had lighter underparts. The area around their mouths was also paler. Adults were about 6.7ft (2m) in length. Newborn seals were approximately half that size.

World of the West Indian monk seal

- Apart from human hunters, the main predators of the West Indian monk seal in the Caribbean waters were sharks. Human hunters took most of them, however, sometimes killing as many as 100 in a single night.

LIVED: prior to 1952
SIZE: 6.7ft (2.1m) long
HABITAT: Caribbean waters
DIET: fish and crustaceans

OTHER DATA: overhunted for their skins and their meat; ventured on to land to breed; scientific name, *Monachus tropicalis; say:* MON-ARK-OOS TROP-EEK-AH-LIS; status: extinct

Back in 1911, it was reported that some local fishermen had butchered most, if not all of the remaining 200 West Indian monk seals in the Gulf of Mexico. The only possible reason was sheer greed as the price these creatures fetched was considerable. Some people believe it is possible that a few may have escaped, however. Since then, there have been occasional claims that a few appear during annual migrations. But even if these accounts are accurate, any surviving population would be tiny.

FACTFILE

The last reported sighting of a West Indian monk seal was in 1952 in an area in the Caribbean between Jamaica and Honduras.

SAVING OTHER MONK SEALS

An aerial survey was carried out in 1972 by scientists who wanted to investigate whether there might be surviving West Indian monk seals in their natural habitat. An expedition also went to look for some in 1980. However neither of these attempts proved fruitful. But all may not be lost for the monk seals because two related species, the Hawaiian monk seal and the Mediterranean monk seal, both once highly endangered, are now protected.

Steller's sea cow

Described as a gentle giant by the scientist who discovered it back in the 18th century, this huge marine mammal must have been remarkable to behold.

It was among the most magnificent of ocean creatures; but, once discovered by sailors, the days of the Steller's sea cow were numbered. The North Pacific was a cold and barren area, and the sea cow provided a ready source of food for mariners. Its flesh, when cooked, was described as extremely tasty, and one dead sea cow could feed as many as 30 men for a month. The hunters butchered them mercilessly.

Harpooned and then stabbed with all sorts of weapons, they were sliced up while still breathing. Their tails must have flapped furiously and they would have fought with their flippers, but to no avail.

SOCIAL CREATURES

The 18th-century German scientist Georg Steller, who had given his name to this creature, was in fact the only naturalist ever to see these gentle and intelligent beasts, known to have risked their lives in the brotherly attempt to pull fellow creatures free.

LIVED: prior to 1741
SIZE: 30ft (9m) long
HABITAT: on Bering Island in the N. Pacific
DIET: fish of various kinds, and seaweed

OTHER DATA: named after Georg Steller, the explorer who discovered it; scientific name, *Hydrodamalis stelleri; say:* HEYE-DROH-DAM-AH-LIS STEL-ER-EE; status: extinct

But by 1768 – just 27 years after Steller first encountered them – these huge and peaceful marine mammals, noble both in appearance and nature from all accounts, had been completely wiped out by Man. Two smaller types of sea cow – the manatee and the dugong – however, still exist today. (You can read about them later in this book.) Both are rare and, it is hoped, neither will meet a similar fate to that of Steller's sea cow.

World of Steller's sea cow

- Discovered when Georg Steller was wrecked on Bering Island back in 1741, Steller's sea cow is known to have been related to the elephant.

Extinct frogs

Over recent years, scientists have noticed that some species of frog seem to be declining in number, or are becoming deformed. Others have disappeared altogether.

The Palestinian painted frog was first identified as a species as recently as 1940. But only 16 years later, it seems to have become extinct. About 3.25ins (8cm) long when mature, the Palestinian painted frog was first identified from only two specimens and two tadpoles. Strangely, it is the only painted frog to have been found living east of the Mediterranean Sea. It also differed from other known types of painted frogs in a few additional ways.

First of all, its front legs were longer. It had distinctive markings too, in the form of red and white blotches and spotted areas on its rust body and grey underparts.

It must have been very disappointing for the person who found the first two young frogs because it turned out they had cannibalistic tendencies!

The bigger of the two, which was a female, must have become hungry or disturbed in some way because it gobbled down the other one. There seems to be no record of what happened to the two tadpoles that had been found at the same time, however.

In 1955, one further specimen was discovered; and from the large size of this female it was noticed that the two previous frogs must have been juveniles. No further Palestinian painted frogs, depicted in the illustration across these two pages, have come to light since that date.

TAKING A GAMBLE

Another species of frog that has become extinct is one that lived in the region of Las Vegas, the gambling capital of the United States. Known as the Vegas Valley Leopard frog, it became rare after natural springs were diverted. But that was not all. Conservationists took a risk when they introduced trout to the area because these ate the frogs' eggs and tadpoles. The last time this frog was seen was back in 1942.

World of the painted frog

- Land reclamation is meant for the good; but when swamps were drained on the Israeli-Syrian border, this led to loss of the painted frog's natural habitat.

DATA: reasons for the extinction and decline of certain frogs include changes in natural habitat, often due to land reclamation; introduction of fish that eat both frogs' eggs and tadpoles; collection by humans; pollution of waters; use of pesticides and fungicides

Sometimes, too, frogs have disappeared because people find them cute and like to collect them. This happened with the golden frog of Panama, where it was found in just one tiny region. Tourists took so many of this species that the golden frog is now hardly ever seen.

Other reasons why some frogs have become scarce could well include the liking some people have for eating frogs' legs, and the use of pesticides.

In 1997, the Australian government even banned use of some herbicides. They could not be used near to water because of the harmful effect on frogs and tadpoles.

There have been reports, too, that fungicides can take their toll on frogs. In the North American state of Minnesota, for instance, it was shown they could stunt tadpole growth and affect the normal sexual development of frogs. The pollution of watercourses with detergents, meanwhile, has sometimes prevented frogs from breathing through their skin. Parasites, such as the trematodes, can also sometimes be to blame for disappearing frogs.

FACTFILE

A collector, M. Costa, who looked after a single specimen, found that the Palestinian painted frog was active only at night.

Extinct snakes

The Round Island boa is believed to have been extinct since 1980. Another boa, meanwhile, the Madagascan variety, is severely endangered. First prize for length and girth must go, however, to the ancient giant boa, shown *left*.

An enormously powerful snake, the giant boa would have been highly dangerous, killing its victims by wrapping its body around them and then tightening its grip until they suffocated. It may indeed be what was once thought to have been a mythical serpent featuring in the legends of the Australian Aborigine people.

Fossilized remains show that it grew up to 16.5ft (5m) in length and that it was extremely chunky. Its central body may even have been as wide as your chest.

FACTFILE

Some snakes have become rare because they are hunted for their skins. In some parts of the world, they are even eaten.

DATA: reasons for the extinction of snakes include dislike of these creatures leading to repeated killing by humans; loss of natural habitat; taking of their skins for the fashion industry; the eating of snakes in some parts of the world; in some societies, their blood is even drunk

LOSS OF HABITAT

We do not know why this giant boa became extinct long ago; but scientists put the extinction of several other snakes down to human interference. The boa from Round Island, near Mauritius in the Indian Ocean, for example, is said to be effectively extinct (just one or two may remain). It lost its natural habitat when rabbits and goats were let loose into the wild on Round Island, only to deprive a large area of any sort of vegetation.

In the West Indies, meanwhile, snakes such as the Jamaican tree snake and the Martinique racer have disappeared due to hunting by humans and attack by the predatory mongoose in some instances.

Sometimes snakes are treated extremely cruelly. They are skinned alive in some parts of the world because the quality of their skin is preserved that way. Then they are sold to the fashion trade. In Thailand, they are killed so people can drink their blood which is thought to give the consumer increased vitality. They are caught for food in some places, too.

But why should we be concerned if some species of snakes become extinct? After all, throughout history they have been a symbol of evil and death. The deadly King cobra is even known to have a bite that can kill humans in just a few minutes as its poison spreads through the body.

Snakes are, in fact,very important for controlling pest species, such as rodents and certain unpleasant insects. Indeed, in parts of the world where some types have been entirely eradicated, such pests have become a serious danger by spreading potentially fatal diseases among the human population.

In the future, it could be that scientists may even find that snakes have medicinal uses. Researchers at the University of Southern California, USA, even suspect that a protein in the venom of the copperhead snake can slow down the growth of a form of cancer. Interestingly, venom from the Brazilian arrowhead viper has been shown to assist in the lowering of blood pressure. Save snakes from extinction and we may save humanity.

World of rare snakes

- Many people dislike snakes, even those that are not poisonous, because they are long and slithery, and have flicking tongues. But some are now very rare, and these require protection in the wild if they are not to become extinct.

Extinct lizards

In 1926, the first live specimens of a giant lizard known as the Komodo dragon ever seen outside Asia were brought to a New York zoo. They were impressive, but nowhere near as large as Megalania had been.

Thought to have been an extinct relative of the Komodo dragon, the Megalania, or giant monitor lizard from what is now Australia, shown *below*, was enormous. In fact it may have extended to 20ft (6m.)

Fossils of Megalania have been dated to show that the creature must have lived at the same time as some of our own early ancestors. No wonder many legends grew up around it! Today's Komodo dragons (named in 1912 after the small Indonesian island where they were first identified), however, are only about half that size.

No one knows why Megalania died out. We can only guess that it may have been exterminated by humans who found it horrific because of its size. Perhaps, like the heroes of literature who went dragon-slaying, those who killed Megalanias were praised for eliminating what was regarded as a fearful beast.

Whatever the truth, now the largest monitor lizard to be found in Australia grows to just 6ft (under 2m) in length and has the scientific name *Varanus giganteus.*

Other lizards have become extinct more recently for a variety of different reasons. Some from the West Indies, for example, became the prey of the mongoose, and of cats, rats, and birds.

DATA: the reasons why some lizards have become extinct include predation by other creatures; failure to breed in captivity; overhunting by humans for sport; destruction of natural habitats; droughts and other natural disasters; unwarranted human fear leading to slaughter

Some that have disappeared, meanwhile, were hunted by humans purely for sport; and one particular species, from a tiny island lying near Guadeloupe, vanished after much of its natural habitat was destroyed in a horrific hurricane.

There are times, too, when building work can affect a reptile population. On a tiny island off Minorca in the Mediterranean, for example, the Ratas Island lizard became extinct in 1950 for that very reason.

FACTFILE

There are Australian fossils proving that the giant monitor lizard must have become extinct less than 20,000 years ago.

LIZARDS FOR LUNCH

Droughts, resulting in loss of vegetation, undoubtedly contributed to the demise of the Cape Verde giant skink. Convicts, banished to one area on these islands in 1833 during a terrible famine, were even forced to turn any surviving lizards into part of their staple diet, so that numbers must have dwindled even further very rapidly.

Determined attempts by scientists to breed some rare species of lizard in captivity have sometimes failed, too.

Extinct giant turtles

The shell of Meiolania, depicted here in the illustrations of one giant tortoise following another, was up to 4ft (1.3m) in length. But there have been other tortoises, too, that reached massive proportions.

What was most unusual about Meiolania was its horned skull, which measured about 2ft (60cm) across.Those horns probably meant it could not withdraw its head into its shell. Fossilized remains have been found on Lord Howe Island, off the coast of New South Wales, Australia, and date from Pleistocene times.

Turtles and tortoises first appeared on Planet Earth about 200,000,000 years ago, and the largest ever to have lived was a sea creature, known as Archelon, which existed over 65 million years ago in Cretaceous times, towards the end of the reign of the dinosaurs. It grew to be 13ft (4m) in length, a truly remarkable size for a turtle.

DELICIOUS FARE

Giant tortoises that have become extinct in more modern times include the giant tortoise of Mauritius, which has not been seen since 1700. It probably died out for the same reason as the Rodriguez giant tortoise, which lived on a nearby Indian Ocean island – that is, overhunting for its meat and shell. One person who tasted it said the flesh was so succulent that, even when served plain, you would think it had been prepared with a rich sauce.

From the early 18th century onwards until this tortoise became extinct 100 years later, ships would even stop at the island of Rodriguez just to take onboard as many of these giant tortoises as they could pack into the holds of their vessels.

DATA: the reasons why various giant turtles have become extinct include overhunting for their shells which have always fetched high prices; humans have also hunted them for their meat, said to be delicious; destruction of their natural habitat has also been to blame

Some giant tortoises from the Galapagos Islands in the Pacific Ocean have also become extinct, even though the very name *Galapagos* is derived from a Spanish term meaning "a place where tortoises thrive." Here, once again, large numbers of the tortoises were killed for their meat from the 17th century onwards. The Charles Island tortoise disappeared in 1876, and the Barrington Island tortoise has not been seen since 1890.

The fate of another Galapagos Island tortoise seems to have taken a different turn, however. Once found on Abingdon Island, it disappeared due to hunting by visiting sailors and whalers. In 1972, however, the scientific world was in for a great surprise.

World of the extinct tortoise

- How the tortoise Meiolania got from Australia to the small South Pacific islands on which it lived remains a mystery because it could not swim.

A live specimen and tracks were discovered. Some years later, droppings were also found that were no more than a few years old. Perhaps other supposedly extinct turtles lurk somewhere, yet to be rediscovered.

FACTFILE

The shell of a tortoise is known as its *carapace*. These differ a lot in size and shape, as well as colour and markings.

Aurochs

Thought to have finally disappeared from our planet around the year 1627, the aurochs featured prominently in the cave paintings of our prehistoric ancestors, as you can see in the illustration shown *right*.

A primitive type of wild ox, this creature was about 6.5ft (2m) in height and had very long horns which often extended to as much as 32ins (81cm) in length. A direct ancestor of the domestic cattle known in Europe today and of the black fighting bulls of Spain, it survived in Poland until the 17th century. Here, the very last of the species died in a reserve that had been set up in the attempt to preserve the species.

> **FACTFILE**
>
> The aurochs was killed for meals of its meat, for its bones from which carvings would be made, and for its hide, used for clothes.

SEEN BY CAESAR

In prehistoric times, the aurochs inhabited most of Europe and Asia, and was extensively hunted by humans. The Roman Emperor Julius Caesar was later to describe them as follows in his work *De Bello Gallico* (a Latin title which, in translation, means *About the Gallic War*):

"They are only a little smaller than elephants and are related to bulls. They are very strong and can run very fast. No one is safe when they are around.

"They cannot be tamed, even when they are young. Anyone who kills a great number proudly displays their horns as proof of this and is greatly honoured. The horns differ from those of our oxen and are greatly sought after. Edged with silver, they make excellent drinking vessels for use at important feasts."

CRUELLY TREATED

But it was not only in Roman times that the aurochs was sought after. From the 16th-century writings of Conrad Gessner, we also know that they were hunted with extreme cruelty throughout the ages. In his own words: *"One individual is made to separate from the rest of the herd and is then hunted by many men and dogs, often for a long time. The aurochs only falls when hit in the chest. While it is still alive, the skin between its horns is ripped off. This is sent, together with the heart and its fresh, salted meat to the king who will sometimes send these remains as a gift to other royalty."*

There have been recent attempts at recreating the aurochs by breeding certain types of cattle together, but none has been wholly successful as yet. However, it could be that this may one day be possible using DNA from aurochs remains.

LIVED: prior to 1627

SIZE: 6.5ft (2m) tall

HABITAT: in most of Europe and Asia

DIET: grass and other vegetation

OTHER DATA: extremely strong; very long horns; cruelly hunted by humans; scientific name, *Bos primigenius*; *say,* BOHS PRIM-IJ-EEN-EE-OOS

Extinct fish

The strange fish pictured here are a species known as *Gemuendina* which existed way back in time and which we are unlikely ever to see the like of again. But a lot more familiar fish have more recently become lost from oceans, rivers, and lakes.

Gemuendina was similar in shape to modern manta rays. As one of the *placoderms*, which means part of its body was armoured against predators, it was flat with broad, rounded shoulder fins, and a long, flexible tail, as you can see. From Devonian times, it also had unusual jaws that could be extended to make it easier for this fish to catch its own prey. No one is sure precisely why this fish, known from its fossilized remains, became extinct. But it perhaps disappeared in one of the early mass extinctions, or died out due to disease, or the advent of large marine predators.

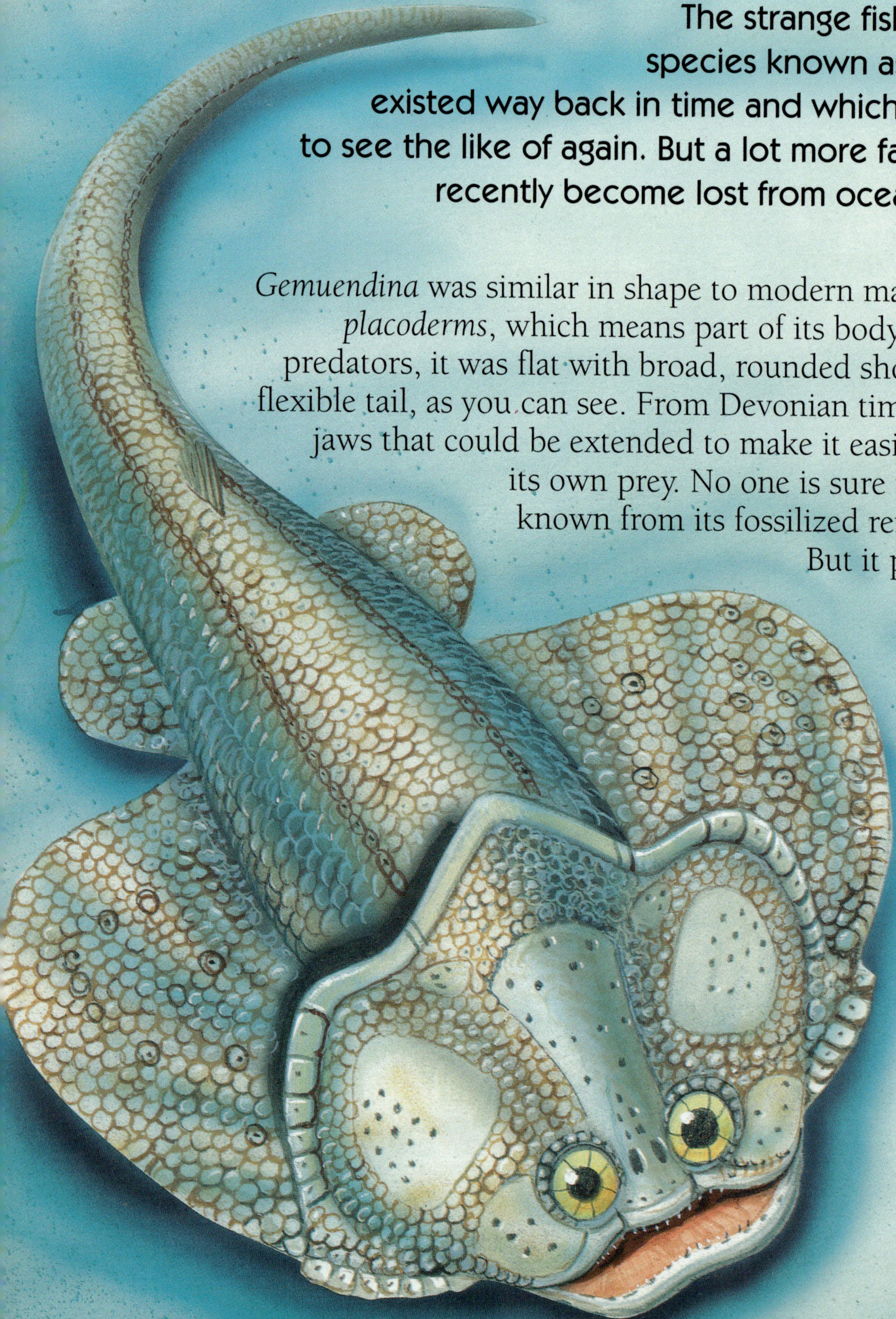

DATA: the reasons why some fish became extinct in prehistoric times included evolving into different forms and possibly predation by sea monsters; more recent extinctions may be due to such factors as the pollution of seas, lakes, and rivers, the draining of some waters, and overfishing

FACTFILE

Genetic tests can be used to identify whether newly found fish in fact belong to a species wrongly declared extinct.

OFF THE MENU!

As this book goes to press, a newspaper report reveals that a growing taste for shark fin soup is putting this sea creature at risk of extinction, too. Environmentalists are therefore demanding that it should be taken off the menu in expensive Chinese restaurants, where it has been a highly regarded dish until now.

It seems, from research carried out by a group known as WildAid, that as many as 100,000,000 sharks and sharklike fish are taken every year for their fins. The problem, however, is principally a financial one. A dorsal shark fin can fetch as much as £10,000 ($15,000 US); and some very rich people are prepared to pay up to £70 ($105 US) for a single bowl.

How sad to think we might lose the shark forever simply for soup!

At the same time, what was once a much more affordable fish is disappearing from Atlantic waters, with the result that certain British restaurants have stopped serving cod in the attempt to revive this species. Stocks in the sea are reported to be badly exhausted due to overfishing, and the European Union's Fisheries Commissioner has stated that drastic measures are needed to guarantee their continued existence.

Overfishing in Canada's maritime provinces has caused a similar dramatic decline. Inland, too, there is concern as the numbers of fish in rivers and lakes decrease year by year. It is also evident that, due to pollution of waters, some fish become deformed in various ways. Somehow we must rectify the situation before it is too late.

Bogus species

If anyone claims to have seen the jackalope in this picture, either recently or in the past, you should accept it with a pinch of salt. Quite simply, the jackalope does not exist, and never has. Instead someone dreamed it up as a practical joke.

In the past, and perhaps even today, either accidentally or in the attempt to hoodwink the public, palaeontolgists and taxidermists have occasionally provided incorrect information. By "inventing" bogus species and then presenting their remains, they have made people think a certain creature once existed but became extinct when, in fact, it is a figment of the imagination. The so-called jackalope in this illustration, for instance, was created purely for amusement by putting a deer's antlers on to the remains of a hare's head and body, which were then stuffed.

DATA: sometimes people have played practical jokes on scientists by altering fossilized remains to create a curious species that never actually existed; but occasionally scientists have made genuine errors when reconstructing extinct creatures, as with some early discoveries of dinosaurs

PILTDOWN MAN

Another well-known example is that of Piltdown Man. Back in 1912, there was great excitement when it was thought that palaeontologists had found the fossilized skull and teeth of a "missing link" between humans and apes. It was even said to have belonged to the most ancient inhabitant of England, and possibly of Europe. The remains had been unearthed in a gravel pit near Piltdown Common in the county of Essex, which is how the creature came to get its name.

Before long, even academics were speculating as to what Piltdown Man must have looked like. They said he must have been short and that he probably ate elephant and rhino meat.

Others were more sceptical, however, and thought the remains were not genuine. An article in a magazine even went so far as to express the view that the remains might have been *"artificially fossilized and planted in the fossil bed to fool the scientists."*

As it turned out, the skeptics were right. But not until 1953 was the skull finally identified as human and no more than 500 years old! It had all been a deliberate hoax. Yet no one knows for certain to this day who it was who started it all.

FACTFILE

Hoaxers have sometimes placed false extinct animal footprints in places. But experts usually are not fooled for long.

Was Charles Dawson, an archaeologist who found the skull, the perpetrator of this joke, or did someone else fool him and countless others? It may always remain a mystery.

MAN OR BEAST?

Sometimes, however, genuine errors are made by the experts and we must forgive their mistakes. An interesting bone first studied in the 17th century, for example, was thought to have belonged to a giant human. It was illustrated but then lost. However, scientists now think it is more likely to have been the lower end of the thigh bone of a large dinosaur, known today as a Megalosaurus. When experts first put together the bones of a plesiosaur, they also made a gross error and put its small head at the end of its tail; and the dinosaur Iguanodon's famous spiked thumb was thought at first to be a horn.

World of bogus species

- The dinosaur Tsintaosaurus, from China, was also thought to have had a horn at first and to have looked like a unicorn, but that "horn" was merely broken bone.

In danger today

Each of the creatures bordering these two pages is classified as endangered. So few of them remain today that conservationists are concerned for their future.

The International Union for Conservation of Nature and Natural Resources regularly publishes what is known as the *Red List*. The colour, here, naturally enough, stands for *Danger*; and the list cites those creatures fast disappearing from our planet. So which are they? And why is their survival currently threatened?

Several of the species you are about to meet have been hunted to near-extinction by humans – either for food, for sport, for clothing, or for status symbols. In other cases, destruction of natural habitat may have occurred. But sometimes extinction is a simple fact of life and part of the evolutionary process. As you look through the final section of this volume, you will meet many of the world's most wondrous creatures – the cheetah, the Florida cougar, the blue whale, the orang-utan, the giant panda, the spectacled bear, the Atlantic walrus, and the Japanese crane among them. In almost every instance, numbers are decreasing with each passing year; but, you will find, it is not all bad news. Read on to find out why!

Giant panda

The Chinese name for the giant panda is "*Da xiong mao,*" which means "great bear cat." For thousands of years it has been venerated by the Chinese and thought to have special powers. It was even kept as a pet by the ancient Chinese emperors to ward off disasters.

According to estimates, there may be as few as 700 giant pandas existing in the wild. Their natural habitat within China comprises areas of damp coniferous forests, where the bamboo shoots, leaves and stems that are their main source of food can be found.

Curiously, though, the giant panda finds it difficult to digest bamboo. Most of this intake therefore passes straight through its body, so it has to take in and chew great quantities using its strong jaw muscles in order to obtain sufficient nourishment. Indeed, if bamboo is the only food source available, the giant panda may spend half the day at its meals! Otherwise, its diet comprises grasses and tubers of various kinds, and the occasional insect.

All this food intake helps the adult maintain its average body weight of about 200 pounds (90kg.)

THE YOUNG ONES

Very few cubs have been born in captivity, so that not much is known about the breeding habits of the giant panda. Scientists believe, however, that the mothers normally give birth to only one or two offspring and that probably just one survives. They are tiny, weighing only a few ounces on average, a few ounces (100g) at birth, and blind until about six weeks old. At birth, they are pure white; but within a month they develop adult colouring.

These cubs are so small that they are extremely vulnerable to predators. It is therefore up to the mother to protect them from attack, and she will do so for about 18 months. By then, the cub has long been weaned, and the mother may be ready to breed once more. Now it is time for her to drive off her cub, and it will have to fend for itself.

World of the giant panda

- Giant pandas were first described by Père Armand David in 1869 when he discovered them in China. Their natural habitat is exclusively the cloud forests in the three Chinese provinces of Sichuan, Kansu and Chensi.

SCIENTIFIC NAME: *Ailuropoda melanoleuca*
say: AL-YOUR-OP-OD-AH MEL-AN-OL-OO-KAH
HABITAT: coniferous forests of China
DATA: cubs are tiny at birth and vulnerable to predators; other dangers to this very rare creature include becoming caught in traps set for other animals, but not heavily hunted; now a conservation symbol; status endangered

FACTFILE

Symbol of the Worldwide Fund for Nature, the giant panda first became known to the western world in 1869.

Aye-aye

SCIENTIFIC NAME: *Daubentonia madagascariensis*
say: DOW-BEN-TOH-NEE-AH MAD-AG-ASK-AR-EE-EN-SIS
HABITAT: forests of Madagascar
DATA: decline due to severe deforestation, and slaughter by humans; the famous British palaeontologist Sir Richard Owen classified the aye-aye as a type of lemur in 1866; status: endangered

There are precious few aye-ayes left in the wild, and they are found only in the north-east coastal forests of Madagascar, which lies off Africa.

Experts think that the destruction of much of the natural forests on Madagascar during the last half of the 20th century probably led to a dramatic decrease in the number of aye-ayes to be found on the island. But even when it was first discovered, it was probably fairly rare. Madagascar has been isolated from the rest of the continent of Africa for many millions of years, and as a result has developed flora and fauna that are unique in many ways.

Pronounced *"eye-eye,"* the aye-aye's name seems very apt! This is because this nocturnal creature's glowing organs of sight are so large and efficient. This insectivore's other main characteristics include its hands, on each of which is a very long and thin third finger. This finger comes in useful for tapping at trees in the search for wood-boring insect larvae which will make a delicious meal for the aye-aye.

As it does this, the aye-aye also listens very carefully with its sensitive ears to assess the exact position of the larvae. It will then use its incisor teeth to break through the timber, and its elongated middle finger can now be employed to extract the grubs.

The aye-aye also uses that long third finger for grooming and will hang suspended from a branch by its back limbs while it does this. That is not all. When drinking water or taking milk from coconuts, it will put the elongated digit into the liquid and use it as a sort of pump to raise the liquid to its mouth.

FALSE BELIEFS

Research has shown that the people of Madagascar (known as the Malagasy) have long regarded the aye-aye with fear, even though it is so small and cute-looking. Somehow a belief arose that this creature was a reincarnation of ancestors, or that the sighting of one meant that death was imminent. Fortunately, such superstitions have largely died out; but in the past the aye-aye would be killed by some of the local population through sheer dread of what it was thought to signify.

World of the aye-aye

- When first discovered in 1780 by the French naturalist Pierre Sonnerat, the aye-aye was thought to be a member of the squirrel family. It took eight more years for it to be identified correctly as a type of lemur, native to the island of Madagascar.

Manatee

Although officially protected, the manatee is still vulnerable to those who continue to kill this large and slow-moving sea creature for food or for its hide, which they use to provide hard-wearing leather.

There are three main types of manatees. As its name clearly suggests, the Amazonian variety inhabits the lakes and rivers of South America's Amazonian Basin. Those of West Africa, and the West Indies and North America, meanwhile, live in the shallow coastal waters and river estuaries to be found in these regions.

Sixty years ago, there were probably about 10,000 manatees in the waters around Florida alone.

World of the manatee

- Manatee calves are large when they are born and can weigh about 40 pounds (18kg) at birth.
- As mammals, manatees suckle their young until they can feed for themselves. Their birth weight increases almost 20 times in their first year! If humans grew so rapidly, they would weigh about 1,400 pounds (435kg) by their first birthday!

In early 2001, as revealed by an aerial survey carried out by the Florida Marine Research Institute, there seemed to be in the region of 3,270 manatees in the region. This figure is not too disappointing, however, as the very fast rate at which the manatee had previously been declining seems to have been halted.

COLLISION COURSE

In addition to being taken by human predators, manatees also often collide with watercraft, possibly due to what scientists think may be their shortsightedness. However, the way in which they manage to find their way around in the dark or in murky waters is not understood. Sometimes they wander far beyond their normal range. Prolonged exposure to cold fronts, admittedly unusual in the warm waters of the manatee's habitat, is can also be dangerous for this marine animal. Red tides may have their effect on the species, too. The animals' very slow rate of reproduction is not helpful either in maintaining the species.

On average, as adults, manatees measure about 10ft (about 3m) in length and have very muscular bodies. Their skin is thick and rough, yet experiments have shown that they react to being touched or tickled.

SCIENTIFIC NAMES: *Trichechus manatus/sengalensis*
say: TRICH-E-KOOS MAN-AT-OOS/SENG-AL-EN-SIS
HABITAT: North America, West Indies, and West Africa; and also the Amazon Basin, South America

DATA: decline in numbers due to hunting by humans; collision with watercraft also sometimes to blame; exposure to cold fronts and red tides are a danger to them, too; status: endangered

Their weight ranges from 1,000 to as much as 3,000 pounds (or about 454-1360kg.)

Manatees spend most of their time eating and, as herbivores, may even gobble up to one-tenth of their body weight each day in vegetation.

FACTFILE

Manatees seem to be able to communicate with sound, and recordings have been made of their squeaks and squeals.

A GREAT SCHEME

Conservationists all over the world do their best to protect all endangered animals; but their efforts require considerable funding, and government aid is frequently inadequate for the work that has to be done. This is why many charities are set up in support of such activities.

The Save the Manatee Club is one such example and invites membership in the attempt to raise money to support conservation of this species.

You can find details on the Internet. One delightful method has been to encourage adoption of a manatee for a loved-one in celebration of St Valentine's Day on February 14th for a modest subscription.

California condor

If everything goes to plan, this truly great bird may soon grace the skies over the west coast of the North American continent once again.

Imagine the scene. A California condor is soaring majestically over a small town below. For the moment the air is still. But panic could arise at any time – for this is 1849, the height of the Californian gold-rush. Thousands have come to the west coast of America in search of their fortune, bringing with them weapons and poison.

Ever keen to fill their pockets, the settlers use their guns for shooting at these giant birds, hoping to sell the stuffed corpses to museums.

FACTFILE

The adult bird has an enormous wingspan which extends to over 9ft (2.7m), and weighs about 23 pounds (10kg.)

They use poison, meanwhile, to kill coyotes, wild dogs that roam the land. But condors sometimes catch the coyotes and in turn are poisoned.

They suffer, too, from eating the lead shot left in deer by hunters. Other small animals are also hunted by settlers; so that, in time, there is less and less prey for the condor. No wonder that this magnificent bird gradually became rarer and rarer. To add to the problem, it only mates once every two years – with the result that, by the mid-1970s, there were just fifty thought to be left in the wild.

SCIENTIFIC NAME: *Gymnogyps californianus*
say: JIM-NOH-JIPS CAL-EE-FORN-EE-AHN-OOS
HABITAT: North America
DATA: enormous wing span; once killed in large numbers for museum specimens; also declined because ate meat of coyotes poisoned by hunters; fifty thought left in wild by 1970s; became a protected species; status: endangered

Many were old and in poor health. Indeed, experts rapidly realised that if something was not done soon, the California condor could vanish completely. Eggs were therefore taken from one of the last remaining nests. These were hatched and nurtured in the San Diego Zoo, in California, USA. Released into the wild, it is hoped that they will continue to breed so that the numbers will gradually be restored. The California condor, may yet soar again in the American skies!

Black rhino

According to the International Rhino Foundation, in the 20 years up to 1992, 96 per cent of the black rhino population died out. Have things improved since then?

The black rhino is a mighty creature, weighing up to 1.5 tons – that's the equivalent of about 20 fully grown adult men. To the shoulder, it measures about 5ft (1.5m.) Nevertheless, it is smaller than the white or square-lipped rhino.

Famed for its two magnificent horns, it sports a prominent larger one at the front and a smaller horn grows closely behind it, as you can see in the illustration shown *opposite*.

It is these horns, which are present in both sexes, that have long been avidly sought by poachers because their supposed aphrodisiac properties have led to high demand internationally. And even though such hunting is banned by law, this sort of activity still continues to some extent.

No one, however, has ever proven that rhino horn does actually work in this way. So it is difficult to understand why people continue to believe it.

FIERCE BEAST

The white rhino has a gentle nature in spite of its size. The black rhino, however, can be highly aggressive. It has a very bad temper; and if you were to disturb it, the chances are it would charge at you at a frantic speed. It is, however, very shortsighted. So how would it know you were nearby? It might smell you. Alternatively, tick birds, which feed from insects on this rhino's back, may cooperate with the fierce beast and repay its hospitality by advising it through warning calls of your imminent approach.

World of the black rhino

- The scientific name for the black rhino is *Diceros bicornis*. The first part of this name, *Di*, comes from the Greek word for "two, " and the second part, *ceros*, means "horn."
- In 1960, there were between 11,000 and 13,5000 black rhinos in Africa. Now there are only about 2,700.

SCIENTIFIC NAME: *Diceros bicornis*
say: DIS-ER-OS BEYE-KORN-IS
HABITAT: Africa
DATA: many died through severe drought; sometimes killed for their hide; also hunted for their flesh; many culled to establish agricultural settlements in 1940s; status: severely endangered in their natural habitat, and in reserves

Père David's deer

In 1939, the last known Père David's deer to exist at that time in the wild in China was shot. However this creature was saved from extinction because a 19th-century emperor had kept a herd of them in his park near what is now Beijing. Some were taken to Europe for breeding.

It was the French missionary Père Armand David who first identified the deer that are named after him when he took a furtive look into the emperor's private park in 1865. The emperor donated a few of the deer to the French ambassador, and some were also sent to England and Germany. It was just as well because, in 1895, most of the Imperial herd was destroyed in dreadful floods; and those that escaped this natural disaster were killed and eaten by the starving population.

Later, eighteen of the deer that existed in Europe were sent to the estate of the Duke of Bedford at Woburn Abbey, England. In spite of many difficulties during war years, one stag and five hinds bred successfully, and finally there were hundreds. Over the years, a number have been taken to Beijing and reintroduced to the very area of the former Imperial park, providing a story with a happy ending and a good example of species conservation.

SCIENTIFIC NAME: *Elaphurus davidianus*
say: EL-AF-YOUR-OOS DAV-ID-EE-AH-NOOS
HABITAT: originally China
DATA: disaster struck when deer that escaped floods killed by starving peasants; numbers dwindled and extinct in China by 1921; but reserve had been set up in England prior to that; some returned to China; status: endangered

Père David's deer (also known by the scientific name *Elaphurus davidianus*) can still be seen at Woburn Park on the Duke of Bedford's estate in southern England. They graze mainly on grass and eat water plants when available. But what makes them truly distinctive among other types of deer is that the tail is tufted and longer, and that sometimes two sets of this deer's antlers grow in the same year.

Prior to breeding, in the rutting season, the stags will spar with each other over mates, rearing up on their hind legs and fighting both with their antlers and their teeth.

This is, however, no more than a form of display and an attempt to prove which is the strongest of the males. The females normally give birth to just one offspring in April or May. This offspring will be mature within 14 months.

World of Père David's deer

- This deer is reported to be a good swimmer, spending a lot of time in the water during the summer months.
- All Père David's deer that currently exist are probably related to those cared for at Woburn Abbey, England.

Przewalski's horse

Now only found in captivity, this rare creature was discovered in the late 19th century by Colonel N. M. Przewalski, after whom it is named. But there are plans to reestablish the horse in the Mongolian plains.

Also known as the *takh* in Mongolian, Przewalski's horse stands only 53ins (135cms) tall at the shoulder and is the only wild horse that exists today. Well over a thousand live in captivity in various parts of the world, however.

The principal characteristics of this horse are an upright mane, clearly visible in the illustration shown here, long ears, and a donkeylike tail with longer hairs at the bottom than at the top. In the wild, they always lived in groups of up to 20 individuals. These groups or herds were of two kinds. One was a family group, comprising a male, perhaps two females, and their young. The other type of group comprised stallions only.

World of Przewalski's horse

- The Foundation for the Preservation and Protection of the Przewalski's horse was founded in 1977 in the Netherlands and aims to reintroduce this endangered species into the Eurasian steppes. It works in close cooperation with the Mongolian Association for Conservation of Nature and Environment. Meanwhile, these horses exist in zoos.

SCIENTIFIC NAME: *Equus przewalskii*
say: EK-WOOS PR-ZEV-AL-SKEE
HABITAT: plains of Mongolia
DATA: sole surviving wild horse; decline due partly to harsh weather; hunting by humans for food and interbreeding also to blame; some successfully bred in zoos and wildlife parks; status: endangered

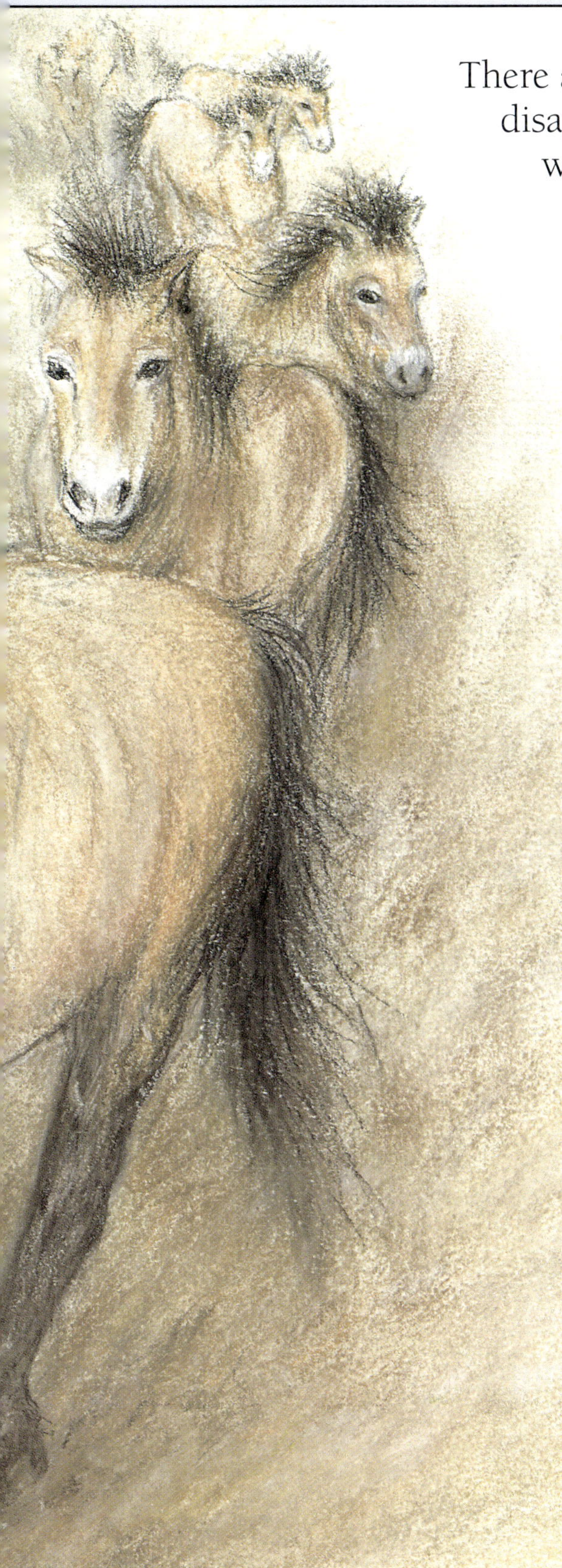

There are several reasons put forward by scientists for the disappearance of Przewalski's horse in the wild. Harsh weather conditions may have been partly to blame. But this horse may also have interbred with domesticated Mongolian ponies. It was almost certainly hunted by humans for food at one time, too. The herds also doubtless met with competition from other animals over grazing lands.

CHANGE OF COAT

In the summer, the coat of a Przewalski's horse lightens to almost a pale yellow shade, and its underparts will be white. In winter, however, the coat thickens and darkens to a yellow-brown. It is, in fact, thought to be a surviving ancestor of the modern domesticated horse; and several similar fossilized skeletons have been found, dating up to Late Pleistocene times. For a considerable time now, there have been more captive stock than exist in the wild; but those dedicated to its survival are doing their best to change this situation.

FACTFILE

Przewalski's horse can run at great speed over a flat landscape, and in the wild has a diet consisting mostly of tough grasses.

Spectacled bear

It is not hard to guess how this type of bear, native to several countries in South America, got its name. Take a look at the area around the eyes of this white-faced mother and cub. Now stretch your imagination. It does look as if they could both be wearing glasses, doesn't it?

The "spectacle" markings around the eyes of this species of bears vary in pattern from one individual creature to another. No one knows for sure why this should be, but it may perhaps help one spectacled bear to recognise another. For the moment, though, the sad truth is that there are less than 2,000 such bears in the wild.

World of the spectacled bear

- Spectacled bears seem to spend most of the day sleeping, usually only coming out at night.

Hunting and destruction of natural habitat have both contributed to the decline of the spectacled bear, but the Spectacled Bear Species Survival Plan aims to rectify the current situation; and this bear is already protected by international law. But they need to act quickly for a successful outcome.

The favoured habitat of the spectacled bear (also known by the scientific name *Tremarctos ornatus*) is the cloud forest of such South American countries as Peru, Bolivia, Ecuador, Colombia, and Venezuela, and at lower elevations of the Andes Mountains. This type of natural environment has widely been destroyed by deforestation for subsistence farming.

SCIENTIFIC NAME: *Tremarctos ornatus*
say: TREM-<u>ARK</u>-TOS ORN-<u>AH</u>-TOOS
HABITAT: South America
DATA: descended from extinct bears common in last Ice Age; only bear in South America; destruction of its natural forest habitat mainly responsible for decline; also hunted for its organs and skin; status: endangered

The spectacled bear has long been among the largest creatures of South America. Weighing an average of 275 pounds (125kg), it is approximately 2.5ft (0.75m) tall when on all-fours, and varies from other bears in that it has 14 pairs of ribs, one pair fewer than they do.

As you can see from the illustration, they have shaggy coats, stocky bodies, and small, pointed ears. For the most part, they are fruit and plant-eaters; but on occasions they will eat insects and small rodents or birds.

When a spectcled bear cub is born, it is very tiny, and may weigh as little as 18 ounces (500g.) Usually only one or two are born at a time between November and February.

FACTFILE

Spectacled bears are sometimes taken by poachers and their body parts used, unnecessarily, as supposed medicines.

Sumatran rhino

The smallest of all the rhinos existing today, the Sumatran rhino is also one of the rarest of the rhinos, found in very small numbers in the rainforests of Burma, Malaysia, Borneo, and Sumatra, of course. Still hunted for its horns, even though this is officially outlawed, it has suffered, too, from destruction of its habitat.

FACTFILE

This rhino's horns, hide, blood, bones and urine were all long thought by some people to have aphrodisiac qualities.

SCIENTIFIC NAME: *Didermocerus sumatrensis*
say: DEED-ER-MOSS-ER-OOS SOOM-AH-TREN-SIS
HABITAT: rainforests of South-East Asia
DATA: destruction of this rhino's natural habitat is a primary reason for its decline; also been hunted for its horns and hide; protected by law but its remains continue to fetch high prices for poachers; status: endangered

Current statistics show that there may be only a few hundred Sumatran rhinos still living in the wild, an extremely low and very worrying figure. Over a period of 10 years, the population went down by 50 per cent. If, therefore, nothing is done to increase the population quickly, it may be that this wonderful creature will have to feature in the extinct section of the next edition of this book. How regrettable that would be!

The scientific name for the Sumatran rhino is *Didernocerus sumatrensis;* and it is known to love to bathe in streams. Indeed, it is a strong swimmer.

GREEDY POACHERS

In the past, laws attempting to protect this rhino did not prove successful in some areas. Poachers were not allowed to kill the creature. However, the sale of its body parts was not prohibited. So poachers were prepared to take considerable risks for the large sums of money that an entire corpse or the small horns and other organs of the Sumatran rhino would fetch.

The Sumatran rhino is strictly an herbivore, living mainly on bamboo plants, figs, other types of fruits, and vegetation. It needs to eat a huge amount in the course of a single day – perhaps as much as 120 pounds (50kg) of food intake. (Amazingly, this is roughly the equivalent of 240 large vegetable burgers!)

Shy and much more gentle creatures than some other species of rhinos, the Sumatran rhinos (also known as the Asiatic two-horned species) live solitary lives most of the time, coming together only when the females are on heat and ready for breeding.

Another Asian species of rhino, the Javan rhino, also became endangered for similar reasons to the Sumatran rhino, except that Javan rhino's horn and blood were always considered even more valuable. Once this single-horned rhino was also found in Thailand, Vietnam, Laos, and parts of India and China, but it is now restricted to one part of Java and Vietnam. According to statistics released in 2000, there were only 68 Javan rhinos surviving in the wild in both places, and none in captivity at that date.

World of the Sumatran rhino

- Two pounds (about 1kg) of rhino horn will currently fetch about £40,000 sterling ($60,000 US) in the Far East. Its main customers come from mainland China, Taiwan, and South Korea.
- A global captive breeding programme is being developed in the attempt to save the species from extinction.

Tiger

On the verge of extinction, this magnificent animal has not only been killed for its body parts but also because people feared possible attack.

This large, colourful animal, readily identifiable because of its stripes, has probably been hunted by humans ever since the Stone Age. Some tigers may have killed cattle and there were occasional human victims. But did this ever justify, in far more recent times, before they became protected to an extent, that some were shot purely for sport? Trophy hunters sought their skins, and these would then be put proudly on display. Thankfully, however, attitudes have widely changed; and most people with some sort of social conscience would be horrified at the thought of a tiger being killed simply to place its head on show.

FACTFILE

The largest and most powerfully built of all tigers is the Siberian variety. It also has the longest coat of any type of tiger.

Destruction of their natural habitat by humans in favour of agricultural land has also been to blame for the decline of many types of tigers, among them the Chinese, Caspian, Indian, Sumatran, and Siberian varieties. There has also been a decline in their principal sources of food through hunting.

MIRACLE CURES?

One of the reasons why the Chinese tiger has become so rare is that many of its body parts have been widely sought after in the belief that they can cure various ailments.

However, there is no real proof of this. An additional problem has also arisen. Dishonest traders have been substituting the organs of other rare creatures, stating that they are genuine Chinese tiger parts, with the result that the survival of other wild cats has become threatened, too.

According to one very dubious belief, if you carry any part of a tiger in your pocket, it will give you courage or protect you against being shot. Curiously, too, the penis of the tiger is regarded as a great delicacy and commands a huge price.

How difficult it is to understand how anyone could be so cruel as to kill such a rare and beautiful creature for these unacceptable reasons!

Probably only a few thousand tigers remain worldwide, both in the wild and in captivity. But *panthera tigris tigris* (its scientific name) is a prime concern to organizations such as Save the Tiger which are working hard to raise funds to protect these magnificent creatures from the fate otherwise awaiting them.

SCIENTIFIC NAME: *Panthera tigris tigris*

say: PAN-THERE-AH TEEG-RIS TEEG-RIS

HABITAT: China, India, Siberia

DATA: disappearance due to loss of forest habitat; taken for its skin and body parts which are thought by some to have use as medicines; decline of natural prey also a problem; status: endangered

Cheetah

The sight of a cheetah on the run is almost unbelievable. It can reach a speed of about 68 miles (110km) per hour in a matter of seconds. A single stride can be as long as 26ft (8m); and every now and then, while it is racing, its feet are all off the ground, just as in this illustration.

With a tan-coloured coat, marked by round black spots, the cheetah is about 4ft (120cm) in length and has a long, fairly thick tail which tends to be held upright behind the animal as it runs at speed in its homeland of Africa.

SHEDDING TEARS?

Perhaps you have noticed that part of the cheetah's body patterning include black marks running from the eyes along the side of the nose, right down to the mouth. It almost makes the animal look as if it is shedding tears for its decline!

Once the cheetah was also found in Asia, but just a few remain there. Why, though, have numbers fallen so dramatically that under 13,000 are thought to survive worldwide? Poaching and destruction of its natural habitat are once more at fault. Cheetahs also fail to breed well in captivity.

FACTFILE

An Egyptian tomb painting, thousands of years old, shows a man wearing a cheetah skin as a sign of his wealth.

SCIENTIFIC NAME: *Trichechus manatus*
say: TREE-CHE-KOOS MAN-AH-TOOS
HABITAT: Africa
DATA: fastest animal on this planet; exquisite body markings; principal cause of decline, hunting by humans for their pelts; also hunted by leopards and lions in natural habitat; fail to breed successfully in captivity; status: endangered

World of the cheetah

- The largest population of cheetahs anywhere in the world is found in Namibia, in southern Africa.
- Over the last 100 years, the cheetah population has drastically declined from the 100, 000 that once existed

ON THE HUNT

Cheetahs capture their prey – antelopes and warthogs, for instance – after stalking and chasing them. Then they usually go for the throat. But cheetahs, too, are hunted by such predators as lions and leopards.

A GREAT MISSION

With headquarters in Namibia, the Cheetah Conservation Fund aims to research and carry out strategies for saving the cheetah, and cooperates with other countries, such as Botswana, Zimbabwe, Kenya, Tanzania, and South Africa, which have smaller cheetah populations. They have also undertaken an education programme among children and farmers, and welcome qualified volunteers for this, too.

Nature has a wonderful way of protecting baby cheetahs, however, from being killed. When they are born, they have long hair along their backs, which they later shed. Scientists believe this hair is useful for camouflaging the cheetah cubs in the grass so that potential predators will not even be aware they are lying there.

Although the swiftest hunter in Africa, the cheetah may not be able to escape further severe decline without the help of such concerned and dedicated scientists in the fight against extinction.

Blue whale

The largest mammal possibly ever to have lived, and certainly the biggest that survives today, the blue whale can grow to 85ft (26m) in length. At present, there are probably just 11,000 of these mighty creatures in the world's oceans, which is why the hunting of this whale is now banned.

Called by its name because of its blue colouring, as you might guess, this baleen whale feeds mostly in the summer when it may eat several tons of krill – comprising many millions of these small crustaceans – in the course of a day.

Like all baleen whales, the blue whale has no teeth. But it has hundreds of special fringed plates which are used to trap food. It also has the most enormous tongue that is the size of a small car. Just imagine that!

At one time, the blue whale was protected simply because of its gigantic size. Hunting was virtually impossible. But once the harpoon gun was invented around the middle of the 19th century, its numbers sadly declined. Each corpse had a high commercial value because of the large quantity of oil it contained.

The problem was that many blue whales lived in waters that did not belong to a particular country. So no one took responsibility, and hunters were not subject to restrictions for many years. How much better it would have been if, back then, the world's many nations had agreed that these waters were owned by everyone and that everyone therefore had a duty to save the blue whale.

SCIENTIFIC NAME: *Balaenoptera musculus*
say: BAL-EE-NOP-TER-AH MOOSK-OO-LOOS
HABITAT: the world's oceans
DATA: can reach high speeds; largest mammal surviving today; once widely hunted; its oil and meat were sought after; even though protected as most countries ban hunting in their seas, still taken by poachers in international waters; status: endangered

In 1930-1931, as many as 29,400 blue whales were killed for their oil, meat and bone meal. Today, less than half that number exist.

Known, too, by the scientific name *Balaenoptera musculus musculus*, the blue whale has usually been found singly or in a pair, but there are also occasional accounts of small schools numbering up to fifty.

BIG BABIES

Blue whales are found in both the northern and southern hemispheres and mate during the winter. Twelve months later, their single calves – weighing an incredible 3 tons at birth – are born. It is hard to believe, too, that a newborn blue whale calf will be up to 27ft (8m) in length! It must be difficult for a blue whale mother to provide sufficient milk for her calves. But the birth rate is slow, and females raise only one offspring every two or three years.

BUOYANT BODIES

If a creature the size of the adult blue whale was land-based, it would need the heaviest bones imaginable to support its great weight. In the sea, however, it is supported by the water, so the absence of massive bones is therefore not important.

Take a careful look at the illustration across these two pages, and you will see how streamlined the blue whale's body is, in spite of its size. The head is rather flat, and its body is usually smooth, except that barnacles sometimes become attached.

The blue whale is an extraordinarily strong swimmer and fantastic speeds can be achieved in spite of its size. It can even match the permitted speed of a car in a city environment if its routine is disturbed in some way.

FACTFILE

Other severely endangered whales include the sperm whale, the grey whale, the humpback whale, and the fin whale.

Wild Bactrian camel

Sadly, the most up-to-date figures show there may be as few as one thousand wild Bactrian camels left on the planet.

In Mongolia and in China, the wild Bactrian camel, one of the world's rarest species, faces many dangers. In Mongolia, they are often attacked by wolves; and in China, they are frequently threatened by poachers, either for food or purely for sport.

We can only hope that conservationist activity will act in time to save these ships of the desert.

Bactrian camels are remarkable creatures. They are herbivores; and although they prefer grass and green shrubs, they will adopt a diet of dry vegetation if nothing else is available. Body fat stored in those two large humps then provides sufficient nourishment and water for a very long period.

They are fast-running animals – speeds of up to 40mph (65km/h) have been recorded – and are also exceedingly strong. Indeed they can carry very heavy loads, perhaps as much as 600 pounds (270kg) and will travel long distances without tiring. They are good climbers, too.

FACTFILE

An easy way to remember the Bactrian camel has two humps is to turn its initial, B, on its side, and you will clearly see two humps!

They can survive in a huge range of temperatures, and have the ability to close their nostrils in a sandstorm so breathing is not impaired and their good sense of smell maintained. Very long eyelashes also protect their efficient organs of sight from such storms. During very hot summers, the few that survive today are thought to climb to a high altitude; but in winter they come down again to the desert regions. In summer, their fleece protects them from the very hot sun, while in winter it keeps them warm. They live mostly in small groups of up to 30 animals, led by an adult male, but some lead a solitary existence.

BIG DRINKERS

What, then, are the other main characteristics of these camels? One extraordinary feature is that, although they can survive for months without drinking water, if it is available to them, they will drink as much as 100 pints (57 litres) at a time! Through necessity, they have also managed to adapt to drinking salt-water slush, something unpalatable to most other animals. If their humps are upright and full, this usually indicates that they have fed well. Sloping humps, however, indicate malnourishment.

SCIENTIFIC NAME: *Camelus bactrianus ferus*
say: KAM-EL-OOS BAK-TREE-AHN-OOS FER-OOS
HABITAT: Mongolia and China
DATA: hunted for centuries for its meat and also for its hide; also taken for use as a pack animal; decline due to competition for water, too; does not breed in captivity; now protected as a species; status: endangered

The wild Bactrian camel went into decline for a number of reasons. Over many years it was mercilessly persecuted by hunters, and its meat taken to be eaten. Its skins, meanwhile, were used for clothing. Sometimes it would be milked. Camels' milk is said to be very nutritious in content and lower in fat than cows' milk.

Young wild Bactrian camels were also widely caught and trained at one time for use as pack animals.

Towards the end of the 19th century, hundreds were sometimes seen together in the Gobi Desert but they were soon to become rare as a species.

As far as we know, no wild Bactrian camel has ever bred in captivity, so there would seem little point in putting them in zoos.

However, a Wild Camel Protection Fund has been set up in Great Britain, and this is working closely with a similar Chinese agency.

American conservationists are also contributing to the establishment of a reserve in the Lop Nor Nature Sanctuary in China. All seek additional funding for their activities.

Every effort is currently being made to save the wild Bactrian camel from extinction. But time is running out; and no one is sure if sufficient funding can be secured to guarantee an increase in the small numbers that still exist. What a tragedy it would be if this noble animal was to die out altogether!

World of the Bactrian camel

- The domesticated Bactrian camel was once the main form of transport on the famous trading route, the Silk Road.
- Bactria, from which the word *Bactrian* comes, was a province in the ancient Persian empire.

Arabian oryx

Conservationists have described the jeep as one of the Arabian oryx's greatest enemies. This is because such vehicles often transported hunting parties who killed just for sport.

The principal features of the Arabian oryx are its two slightly curving horns. Although there are two of these horns on the oryx, it is these, in fact, that led to the legend of the unicorn, a creature with the body of a horse and a single spiral horn.

There are other fables, too, connected with this antelope. Bedouins, for example, once believed that anyone catching an oryx would be sure to inherit its strength. Its horns were also widely prized, and it was hunted as food and for hide. Capturing and shooting an oryx even brought distinction to poachers.

SCIENTIFIC NAME: *Oryx leucoryx*
say: OR-IKS LOO-KOR-IKS
HABITAT: currently only Oman
DATA: poaching is the prime cause of disappearance in the wild; killed for sport; hunting continues today; this is in spite of a conservation scheme known as Operation Oryx; status: endangered

In a description of an oryx hunt in the book *A Look at Threatened Species*, author Lee Merriam Talbot wrote:

"Repeating shotguns are used more than rifles, and often the animals are run until they drop from exhaustion and their throats are cut by servants. It is hard to see how any animal can survive this attack."

But jeeps have sometimes followed whole herds of the antelope for scientific reasons – in order to find out more about the behaviour of this handsome animal. One group of researchers, for example, tracked a herd for 18 hours and found that the animals travelled as far as 58 miles (93km) in that time.

Once, the Arabian oryx was found almost all over the Arabian Peninsula, as well as on land that now forms Jordan and the state of Israel. But the last surviving one in the wild is said to have been shot in 1972. The creature had indeed been severely overhunted over the years if the population had dwindled to one.

Talbot, however, saw a straightforward solution. In his opinion:

"The only way to assure survival of this interesting species is to transfer some specimens to a safe habitat...

"This should be done as soon as possible to be assured of finding enough animals."

FACTFILE

The Arabian oryx has always been able to detect rainfall even from a distance, and will travel towards it to find feeding grounds.

DESERT SANCTUARY

As a result of Operation Oryx, started back in 1962, run by a conservation society known as Fauna and Flora International, and with the full support of a number of Arab countries and various ecological bodies, descendants of a few oryxes rescued for a captive breeding programme were eventually reintroduced into the Arabian Desert within Oman.

This area has been designated the Arabian Oryx Sanctuary, and it is now home to the only non-captive herd of this species in the whole world.

But the oryx's future might not be entirely secure because unless the original cause of the creature's decline in the wild is addressed, there will always be a risk that disaster may strike again.

Poachers have continued to take many oryxes, particularly the females and the young which are extremely vulnerable. As many as 400 Arabian oryxes had been reintroduced from the captive breeding programme but, within three years, there were only about 100 surviving. It is vital that the oryx's human enemies are forced to cease such activities, otherwise all the efforts made to restore the animal to the wild may prove to have been in vain.

Spanish lynx

Once found over most of the Iberian peninsula, the Spanish lynx is now rarely seen. Indeed, there may now be only about 1,400 remaining, according to one estimate. Although it first became a protected species in 1973, it is still in decline.

The Spanish lynx first started to go into decline about 40 years ago when the rabbit population of this corner of Europe – the lynx's main source of food – started to decrease due to a widespread break-out of the disease known as myxomatosis. Poison baits set for other creatures and road accidents have also contributed to decimation of the population. Now about 200 of those that remain are thought to be found in the relatively unspoiled Coto Donana region, where they are protected within a nature reserve.

Among the world's rarest mammals, and nocturnal hunters in the main, these lynxes weigh up to 27 pounds (13kg), and are a golden brown colour, with dark brown or black markings.

Their preferred habitat comprises woodlands and thickets, but they will come out into pastures to hunt their prey. In the absence of rabbits, they will venture several miles in search of a meal, and deer and waterfowl may become their quarry.

STRATEGIES

So what exactly is being done in an attempt to reverse the decline? The Spanish government is pleased to be actively involved in an important conservation programme which outlaws the catching of rabbits in traps in certain areas. Steps are also being taken to increase the number of rabbits in the wild so that there will be more food for the lynxes, while scientists are attempting to breed the Spanish lynx in captivity, then releasing the offspring into the wild.

In addition, there is currently a large fine for anyone caught intentionally killing one of these beautiful members of the cat family. As the result of such determined efforts, may future generations of the Spanish lynx thrive!

World of the Spanish lynx

- These lynxes are sometimes injured or killed by accident in traps that have been set to catch rabbits.
- These severely endangered creatures are more at risk of extinction than any other member of the cat family.

SCIENTIFIC NAME: *Felis lynx pardina*
say: FEEL-IS LINKS PARD-EE-NAH
HABITAT: woodlands in Spain
DATA: disease among rabbits, their main source of food, starved them; too; severely endangered now; protected within nature reserve, however; attempts being made to breed these lynxes in captivity; status: endangered

Orang-utan

Now that so many trees have been destroyed in Sumatra and Borneo, where will orang-utans, the only large apes in Asia, find a refuge? Forest fires, the demand for their meat, and the sale of their babies as pets have led to an even greater decrease in their numbers.

SCIENTIFIC NAME: *Pongo pygmaeus*
say: pon-goh pig-may-oos
HABITAT: forests of Borneo and Sumatra
DATA: decline due to hunting for their meat, deforestation, and trade in babies for pets; mothers killed before the young taken; mothers' skulls often collected and decorated for tourists; status: endangered

If a baby orang-utan is bought as a pet, the odds are that it will not survive for very long. In captivity, they are prone to catching human infections; and the sort of food they are usually given is so different from their natural diet that many become malnourished and starve to death. Others simply become "stir crazy" and die from boredom.

Those that have been forced out of their natural forest habitat into a different, perhaps mountainous area, meanwhile, also experience problems in obtaining the sort of food they need in their new surroundings. In the wild, orang-utans like to eat fruit, leaves, and insects such as bees and termites.

This ginger-haired ape does not flinch when approached by human beings. This may be endearing but it has a big disadvantage. Hunters can get very near to them and then shoot at close range. A current estimation of the entire world orang-utan population is as low as fifteen thousand. What is even worse, they could be declining by as many as one thousand every year.

Babies have often been captured in the wild in very cruel ways. The mother is shot while holding her young, and the baby is then taken from her arms.

FACTFILE

Way back in Pleistocene times, orang-utans were found in parts of India and China, as well as in Java.

One particularly successful conservation programme, however, has an interesting approach. Classing itself as a rehabilitation centre, it has taken in young orang-utans that had previously been in captivity. The task it sets itself is to train these orang-utans to survive independently in a forest environment. Some juveniles formerly in captivity, like the one in the illustration *opposite*, have now even mated with those that had remained in the wild. But there is a further difficulty. Female orang-utans usually only produce one baby at a time, and just every four years. Added to that is the fact that only just over half of all orang-utans born live to reach adulthood.

SWINGING ALONG

A fully grown male orang-utan stands about 3.3ft (1m) tall and weighs approximately the same as an average adult man. Female orang-utans, however, are much smaller. Known by the scientific name *Pongo pygmaeus*, they have very long arms. In fact, if your arms were so long they would reach almost right down to your ankles when you are standing up! These very long front limbs are, of course, very well suited to an animal that loves swinging in the treetops.

Leadbeater's possum

SCIENTIFIC NAME: *Gymnobelideus leadbeateri*
say: JIM-NOH-BEL-EE-DAY-OOS LED-BEET-ER-EE
HABITAT: Australian forests
DATA: always rare as probably prey of many animals; logging decreased numbers further; bush fires destroyed trees where like to nest; once declared extinct but a few found again; emblem of Australian state of Victoria; status: endangered

Now found only in small numbers, and in just a small area of forest in the state of Victoria, Australia, Leadbeater's possums were "rediscovered" in the early 1960s.

One of the main factors affecting survival of the Leadbeater's possum has been the destruction of the type and age of trees, due to serious bushfires and logging, in which they like to nest.

Conservationists are doing what they can to discourage deforestation of the mountain ash; but experts doubt that there will be sufficient nesting sites in old trees to encourage population regrowth for another 150 years. So they are trying to introduce them to artificial nesting boxes, and are keen to find out which type of box these possums respond to best.

First described back in 1867, Leadbeater's possum always has been rare. In fact, by 1909, only 5 specimens had been found. After that, the species was thought to be extinct and declared probably so in 1960.

Then, one year later, a few were found at Tommy's Bend in the Central Highlands of the state of Victoria.

STATE EMBLEM

In Australia, where they are the animal emblem of the state of Victoria, from what naturalists can tell, these possums have always lived naturally in colonies. They are nocturnal, sleeping by day and feeding by night on a whole variety of insects, and also spiders. The exact spots where they can be found, meanwhile, are kept very much a secret so that they will not be unduly disturbed.

But why was Leadbeater's possum so rare right from the time it was first identified? Some scientists think it may have been preyed upon by large monitor lizards, owls, carnivorous marsupials, and feral cats over a very long period of time. It also seems to favour a very particular sort of diet, and in captivity has refused many sorts of fruit, seeds, and leaves. It may also have had to compete with other types of possums, and not fared too well.

World of the Leadbeater's possum

- This species has a very bushy tail which constitutes a large part of its total body length of about 13ins (33cm.)
- It is principally an arboreal creature, and is very skilled at leaping to the branches of neighbouring trees.

Mountain gorilla

The mature male mountain gorilla in the illustration *opposite* has started to develop some grey hairs. Once he starts to do this, he becomes known as a silverback.

Hunters and poachers (dealing in animal body parts), natural disasters, destruction of their natural habitat, and political uprisings – all have had a drastic effect on the mountain gorilla population of Rwanda, Zaire, and Uganda. They are the rarest of all gorillas, and only about 600 are thought to exist at the present time.

Just over half of these inhabit the jungle slopes of the Virunga Mountains, further decline having been halted by the dedicated efforts of such conservationists as the late Dian Fossey. The rest are to be found in an area known as the Bwindi Impenetrable Forest National Park in Uganda.

Gorilla gorilla (yes, twice) *beringei*, to give this creature its scientific name, is extraordinarily shy and feeds mostly on vegetation.

However, this gentle giant will sometimes supplement its diet with some most unusual foodstuff – tree bark, galls, and even the dung of its own species, for instance! In all, it probably spends about one-third of each day feeding – that's as long as 8 hours!

BODY TALK

Male mountain gorillas are much larger than the females, and may weigh as much as 440 pounds (200kg.) Their coats vary in colour, and may range from a bluish-black to brown. A typical adult male mountain gorilla could be over 6ft (about 1.8cm) tall.

Their heads are extremely large in proportion to their bodies, and their arms are long; their legs, short.

The largest male in a social group may mate with a number of females, and they have been observed grooming each other. They are also known to be highly intelligent and capable of learning. Should danger threaten, they give off a strong smell from a gland in their armpits. This, however, has not deterred their main enemy, the human hunter.

World of the mountain gorilla

- Mountain gorilla babies develop about twice as fast as human babies but weigh less than a human child when born.
- Many baby mountain gorillas die before their first birthday, and approximately 50 per cent fail to survive to adulthood.

SCIENTIFIC NAME: *Gorilla gorilla beringei*

say: GOR-ILL-AH GOR-ILL-AH BER-IN-JEYE

HABITAT: forests of Rwanda, Zaire, and Uganda in Africa

DATA: very few exist in wild due to poaching, natural disasters, deforestation, and warfare in these parts of Africa; continual efforts made by conservation schemes to preserve their habitat; status: endangered

Orinoco crocodile

The Orinoco crocodile from South America is the most endangered of all crocodiles and lives mainly in the area of the Orinoco River in Venezuela, as its name suggests. Watch out if one ever approaches!

The prime cause of a dramatic decrease in the number of Orinoco crocodiles surviving today is undoubtedly overhunting for its skin. Countless pairs of shoes, handbags, and belts have been made from its leather.

Even today, in spite of widespread publicity and a whole range of measures designed to protect this species, they are still killed for their skin, for their eggs, and for their meat. Crocodile teeth are also thought by some South American people to ward off evil spirits; and the crocodile's penis, if mixed with alcohol, is said to be an aphrodisiac.

The destruction of their natural habitat does not help matters either. Young Orinoco crocodiles are also sometimes taken and sold as pets. A problem then arises when they grow too big for safety and are killed by their owners.

SCIENTIFIC NAME: *Crocodylus intermedius*

say: KROK-OH-DEE-LOOS INT-ERR-MEED-EE-OOS

HABITAT: mainly Orinoco river in Venezuela

DATA: main cause of decline, hunting for its skin for the fashion industry; eggs taken by birds of prey, too; destruction of natural habitat also a problem; once killed by locals for meat and because teeth thought to deter evil; status: endangered

World of the Orinoco crocodile

- There are three variations in colour of this species of crocodile. Some are dark grey; some may be tan with darker marking; and some may be a greenish grey. Scientists think that some may also change colour when in captivity.

The adult Orinoco crocodile likes to eat fish, all sorts of vertebrates (aquatic and terrestrial), and birds, feeding mostly from the water's edge. There are sometimes reports of them taking humans as their prey, too.

About 16.5ft (5m) in length, this long-jawed, sharp-toothed species is so aggressive that, when observed by naturalists, the males were frequently far too busy fighting each other to take much interest in finding and mating with a female! But on the plus side, it seems that the longer these crocodiles live – perhaps up to the age of 80 years – the more fertile they become, unlike most other creatures, including humans.

The females also lay a large number of eggs – sometimes as many as 70 at a time. This is, of course, a bonus for any conservation efforts. However, in their natural habitat, there is always a risk that the eggs of the Orinoco crocodile will be taken by birds of prey or lizards.

FACTFILE

Only very rarely have there been reports of the Orinoco crocodile attacking humans in recent years.

A MATTER OF BALANCE

Captive breeding programmes are underway, the aim being to increase the number (possibly under one thousand) that exist in Venezuela today. Towards the beginning of the last century, there may have been as many as one million.

Such schemes involve the raising of baby crocodiles in pens after artificial incubation of their eggs. Once hatched, they can be released after 12 months or so, and a few thousand have been reintroduced to the wild in this way.

But what would happen if numbers increased too greatly? What could be done to control such a situation? Once a certain number had been released into the wild, then any captive breeding programmes could be swiftly halted before rivers were overrun by these ferocious beasts. So there should be no need to worry on this score. At least one man, meanwhile, has said he no longer feels safe water-skiing on the Orinoco river!

Dugong

Some people believe that the many myths associated with mermaids are actually based upon sightings of the elusive dugong. Lots of these marine creatures often perish when caught accidentally by fishermen.

In some parts of the world, dugongs are still cruelly harpooned for their flesh, and reported to taste like a combination of pork and beef. They are also prized for their hide, from which sandals may be made. Some become trapped, perhaps accidentally and along with their calves, in nets. Others are caught for their oil and tusks, thought by some peoples of the world to be useful as aphrodisiacs and for the treatment of certain ailments, including headaches and constipation.

Once they liked to live in herds; but as their numbers have declined, they have started to live in smaller groups. They live in warm, shallow seas and never venture on to land. Dugongs need to come up for air, however; and it is this, when suckling their young, that may have led to confusion with mermaids – legendary creatures, said to be half fish and half humans.

BIG AND BULKY

Their calves, too, are born swimmers, and suckle for up to two years. The young are said to bleat like lambs, but the adults are silent. About 10ft (3m) long when fully grown and weighing about 375 pounds (170kg), they are very bulky creatures with a lot of body fat, their colouring ranging from blue-grey to brown.

Other features include a powerful whale-like tail and small flippers. In the males, there are also two large incisor teeth that closely resemble tusks. These seem to continue to grow throughout their lifetime.

If a dugong is left to live its normal span, it will probably survive for as long as a human. The problem is that it is a slow-moving creature and has little by way of natural defence against predatory creatures, such as the humans who have been its principal enemy.

World of the dugong

- The dugong (known by scientists as *Dugong dugon*) is sometimes called a sea cow because it grazes on sea grass.
- These marine mammals are now found almost exclusively around Australia, but a few exist in the region of the Red Sea.

SCIENTIFIC NAME: *Dugong dugon*
say: DOO-GONG DOO-GON
HABITAT: now only in region of Australia and the Red Sea

DATA: also called a sea cow; still caught for meat; tusks also sought after because thought by some to have medicinal properties; sometimes become caught up in fishing nets; status: endangered

Green turtle

You might find it strange that the green turtles illustrated here are a shade of brown, but in fact they vary in colour quite a lot, in spite of their name, ranging from black to brown, grey, and green. Highly sought after, alas, for their shells and meat, they are the largest of all the marine turtles.

World of the green turtle

- The green turtles that make their nests on Ascension Island have been known to swim all the way to the Brazilian coast to find suitable feeding grounds. It is a very long journey even for these very strong swimmers.

Once female green turtles have laid their eggs on a tropical shore, they will soon return to the sea. Scientists trying to study how to prevent them from becoming even more endangered, meanwhile, may sometimes attach radio transmitters to their backs. Then they can track the turtles' movements.

Aquatic tortoises, known scientifically as *Chelonia mydas*, have existed for millions of years, making their current severe decline all the more lamentable.

SCIENTIFIC NAME: *Chelonia mydas*
say: CHEL-OHN-EE-AH MEYE-DAS
HABITAT: in region of Ascension Island
DATA: first existed millions of years ago; tiny when born; hunted by humans for shells and meat; turtle soup once considered a delicacy; many marine predators; status: endangered

When mature, these turtles sometimes weigh as much as 330 pounds (approximately150kg), and might measure up to 3.3ft (1m) in length. But even though large as adults, baby turtles are tiny when they are born, and their carapace may be only 2ins (5cm) in length so that you could hold one in the palm of your hand.

Scientists have made an interesting discovery recently about the sex of these turtles. Those eggs incubated *below* a certain temperature seem always to hatch into males.

FACTFILE

The meat of the green turtle is sometimes used to make soup, once traditionally served at the Lord Mayor of London's banquet.

They leave the beach where they hatched and instinctively, at night, make for the ocean when they are still very small. They therefore risk being eaten there by all sorts of greedy predators, and only comparatively few of a clutch will survive to reach adulthood. Those that do escape are likely to return to the very beach where they were born in order to give birth to the next generation.

Unfortunately, many adult green turtles have been found to have tumours which may decrease the current population even further.

Florida cougar

Although named after the state of Florida, this cougar once roamed throughout other parts of North America, including South Carolina, Louisiana, and Arkansas.

This handsome animal is severely in decline due to slaughter by humans anxious to protect their livestock. It all started in the 19th century when settlers tried to protect their horses in particular from attack, and would give chase to the cougar with teams of dogs. As shown in this illustration, the cougar would seek refuge in a tree and the dogs would ensure it remained there. The hunters would then shoot it. They could not fail to hit such an obvious sitting target.

The cougar population quickly decreased, and the few that remained met with a food shortage as the deer in this part of the United States decreased in numbers, too.

A few cougars survived in an area known as the Everglades National Park, but they have always remained very scarce in that part of Florida, too.

Known by various names in different regions – puma, panther and mountain lion, for instance – the cougar is now fully protected in parts of Florida, such as the Everglades. So few remain, however, that there is little chance of seeing one if you visit a reserve.

SCIENTIFIC NAME: *Felis concolor coryi*

say: FEEL-IS KON-KOL-OR KOR-EE-EE

HABITAT: North America

DATA: known as the puma or mountain lion; would prey on livestock; hunting led to decrease in numbers; environmental pollution and destruction of natural habitat also to blame; status: endangered

Throughout Florida, hunting has long been a popular sport among some of the citizens of and visitors to this state, and hundreds of millions of dollars are contributed to the economy in this way. Deer and hogs are the animals most frequently shot on private land, and this has in turn led to a shortage of prey for the Florida cougar.

Humans have probably always been the Florida cougar's main enemy, while some have disappeared due to becoming the victims of road accidents, the effects of environmental pollution, and destruction of their natural habitat.

But there are several other factors, too, that have contributed to the Florida cougar becoming an endangered species. Scientists think, for example, that once its population had started to decrease, the Florida cougar began to inbreed. This was due to a shortage in choice of mate.

FACTFILE

Large bounties were paid for slaughtered Florida cougars until the end of the 19th century, which doubtless led to many being killed.

INHERITED WEAKNESSES

As often happens with inbreeding, this led to general ill health, heart trouble, and other abnormalities in some of the offspring. A significant number, meanwhile, have developed a highly contagious form of feline distemper, respiratory disorders, and even a certain type of AIDS virus that can attack members of the cat family.

When it was announced that an attempt would be made to introduce into the wild a very close relation of the Florida cougar, there were a number of complaints, even though by nature these creatures are wary of human contact. However, because some had been raised in captivity, they were sufficiently confident to approach people who, in turn, through ignorance, became very fearful, even though there are no authenticated cases of a fatal attack on a human being by this creature. As a result, a few of them were shot, and one got caught in a snare. A more viable method of saving this species has yet to be found.

Polar bear

When confronted by hunters who shoot from the air in spite of regulations that authorities try to enforce, a poor polar bear and her cub do not stand much of a chance. Death is imminent.

Known scientifically as *Ursus maritimus*, meaning "sea bear," polar bears are found only in the Arctic; even though well adapted for survival in Antarctica, you would never find them in the region of the South Pole.

Once again, it is incessant hunting by humans that has led to a steady decrease in numbers. The polar bear's hide fetches a huge price, and hundreds are probably still killed every year even though the total population may be as low as 22,000.

But there are other reasons, too, why the polar bear is in serious decline and at risk of extinction.

In the Arctic regions of Canada, for example, the polar ice starts to melt three weeks earlier in spring than it did only 20 years ago. As a result, the bears have a shorter time to hunt for seals – their staple diet – from their ice platforms. Scientists have even noticed a decrease in their size and the birth rate which they blame on global warming. There are signs, too, that pollutants are taking their toll.

BIG AND BRAINY

Polar bears number among the world's most fascinating creatures. They are massive meat-eaters, and the males stand about 9ft (2.7m) tall – that's a great deal bigger than an average adult human. Their coats are thick, and range in colour from white to a creamy-brown; but much of this variation in shade depends on the light.

Scientist Alison Ames, as a result of studying the behaviour of the polar bear in British zoos, has reached the conclusion that they are very intelligent creatures. She has even watched them break open an ice block to get hold of a frozen fish inside it.

World of the polar bear

- Polar bears are strong swimmers and have been known to swim without stopping for about 60 miles (98km.)
- Sadly, six out of every ten cubs perish during their first year of life. Some die from lack of food or meet with fatal accidents. Others are killed by human trophy hunters.

SCIENTIFIC NAME: *Thalarctos maritimus*

say: THAL-ARK-TOS MAR-IT-EEM-OOS

HABITAT: Arctic regions

DATA: good swimmers; meat-eaters; intelligent; very large in size; food shortages, hunting by humans, global warming, and pollution leading to dramatic decrease in numbers; status: endangered

FACTFILE

Amazing but true – polar bears at San Diego Zoo turned green when algae started growing in their hair shafts!

Atlantic walrus

According to a recent estimate, there are probably about 200,000 Pacific walruses in existence, but as few as 20,000 of the Atlantic species. The decline of the Atlantic walrus was originally due to commercial hunting for its skins, tusks, and blubber.

Walruses have a low reproductive rate. So excessive hunting, as well as global warming, mercury and lead poisoning, reductions in food supply, and frequent disturbance of their natural habitat for oil and gas exploration may well have put the species at serious risk. Saddest of all are reports that when a group of walruses rushes to escape some predator or natural disaster, those in the centre are sometimes crushed to death by the others in the panic.

In Alaska, natives may only hunt the Atlantic walrus for food or souvenir handicrafts from which they earn a living. Carved walrus tusk ivory is still in demand.

Particularly disturbing are companies with licences to hunt for the Atlantic walrus and who offer the opportunity to clients. For as much as £4,500 sterling ($6,000 US), they will take a customer to look for walruses with a view to killing them.

The hunting is done from a motorized boat, also used for fishing by the local Inuit tribe. The organizers recommend suitable rifles and ammunition but advise that the guide will have a harpoon in case the walrus becomes wounded.

They state, however, that it is illegal to import walrus tusks and skins into the United States, and a number of other countries; so, unless they are smuggled in, what happens to the unfortunate walruses that are shot by such trophy-hunters remains unclear.

The walrus also has natural enemies, such as the polar bears and killer whales that share their waters. When human greed and the dangers inherent in a polluted environment become additional risks, the future of the Atlantic walrus is certainly at peril.

World of the Atlantic walrus

- With the scientific name *Odobenus rosmarus rosmarus* (yes, the last part of the name is repeated), meaning "tooth walking sea-horse," the Atlantic walrus is readily identifiable because of its large, very prominent tusks.

SCIENTIFIC NAME: *Odobenus rosmarus rosmarus*

say: OHD-OH-BEEN-OOS ROZ-MAR-OOS ROZ-MAR-OOS

HABITAT: Atlantic waters

DATA: in spite of conservation measures, victims of overhunting by humans; sometimes killed by polar bears and whales, too; pollution of the ocean and global warming also put these walruses at risk; status: endangered

Endangered plants

There are several reasons why plants sometimes become endangered or even extinct. But humans are often to blame simply because, thoughtlessly, they destroy an environment or pick wild flowers without considering how rare these may in fact be. What can be done to save our plants?

The red plant in the illustration *below* is a rafflesia, the largest flower in the world and extremely rare. Found primarily in Sumatra and Borneo, it blooms for only about six days, growing to over 3ft (1m) across. In addition to its enormous size, there is something most peculiar about the plant – its disgusting smell of rotting meat! This has a purpose, however, attracting flies which will then pollinate the plant. Recently, a special cultivation technique has been developed, however, and the rafflesia may yet be saved.

FACTFILE

Wild orchids in particular fetch high prices and are very much at risk of disappearing because of overcollection.

DATA: many wild plants now endangered because they are picked by collectors, because there are changes to their habitat, or due to pollution by chemicals; it is important to try to preserve rare plants in their natural habitats as many do not survive well in an artificial environment

In Great Britain, meanwhile, a brilliant new scheme, known as the Millennium Seed Bank, has been introduced. Launched in the year 2000, hence its name, the Bank has been set up by professional botanists and is described as "a Noah's ark for plants." All over the world, as natural habitats are destroyed or built upon, a great deal of plant life is under threat. Some types of vegetation, having evolved over millions of years, have even disappeared altogether recently.

But seeds, so it seems, have a very long shelf-life. Indeed, some that were hundreds of years old have been known to germinate even though not kept frozen. So prospects look good for the seeds that these scientists plan to preserve at very low temperatures. Within 20 years, the aim is to have in this bank about 20 per cent of all known plants.

The venture, costing millions, has been sponsored by industry. It is, clearly, a very worthwhile investment if, in this way, we can ensure the rescue and subsequent reintroduction of a great many "lost" plants to the planet's ecosystem.

Japanese crane

Officially designated a natural monument in Japan since 1935, this exquisite bird – the rarest of all the cranes – has suffered because of the disappearance of its unique marshland habitat.

FACTFILE

The mating dance of the Japanese cranes is truly spectacular and takes place during the spring.

About 5ft (1.6m) in height and with a magnificent 7ft (2.1m) wingspan, the Japanese crane is famed for its sheer elegance and beauty, and over the centuries has inspired many oriental artists and poets.

However, though protected by the noble classes until the mid 19th century, its numbers began to fall due to hunting and draining of the bird's preferred wetland environment to provide land more suited to human use.

A few once lived in Korea, China, Siberia, Manchuria, and the former Soviet Union, but in these regions they are now severely endangered, if not already extinct. But some have survived in a small remote area of marshy plains on the Japanese island of Hokkaido. Here, in 1924, there were thought to be only 20 of these birds; but this figure subsequently increased due to careful conservation and the creation of a refuge in these wetlands.

Temperatures here are very low in winter, but the Japanese crane can withstand them, and even seems to enjoy the snow.

SCIENTIFIC NAME: *Grus japonensis*

say: GROOS JAP-ON-EN-SIS

HABITAT: mostly Hokkaido, Japan

DATA: performs a spectacular mating dance; draining of wetlands, hunting, and decrease in food supplies led to far fewer numbers, but careful conservation measures now resulting in increased population; status: endangered

Several years of patient attempts at protecting the species paid off, and there is also now a reserve in the south of the island of Kyushu. There are several Japanese cranes at European zoos, too; and some have even been bred there in captivity.

Also known as the Tancho, or Red-crowned crane, the Japanese crane subsists mostly on wetland plants, insects, and amphibians in its natural habitat. But with the draining of the marshes, such food sources have become increasingly scarce.

For a bird that needs large quantities of food to help it survive in inclement conditions, this will inevitably continue to be a major problem.

World of the Japanese crane

- According to Japanese folklore, if this crane approaches someone, he or she will have good luck. It is also known as the bird of happiness.
- Male and female Japanese cranes both share the task of incubating their eggs for about one month after they have been laid.

Puffin

In some places, the puffin population has been declining fast. Oil spills, overhunting by humans for food and feathers, entanglement in nets, and competition for fish are among the factors to blame.

When colonies of birds are seen to be at risk, conservationists will hurry to do all they can to reestablish them. This is precisely what the National Audubon Society (named after the famous naturalist John James Audubon) did back in 1973. Having heard that puffins were becoming rare in the American state of Maine, they set up Project Puffin, a venture that proceeded with considerable success.

Puffins had long nested at a place called Eastern Egg Rock, but hunters killed the last of them towards the end of the 19th century. Scientists, led by Dr Steven Kress, now reacted by taking 100 young puffins each year for 10 years by air, sea and car from Great Island in Newfoundland, which has always had a huge population of puffins. They were then reared artificially on Eastern Egg Rock. The birds were also tagged for identification so that, once released, it would be clear if they did return to this area to nest at the appropriate season. Attempts were also made to lure them there.

This was done by setting up mirrors to give the impression that there were many more puffins at Eastern Egg Rock than there actually were. Tape recordings of the puffin "growl" were also used as a decoy. In the end, a fair number did return to form viable colonies.

FACTFILE

The white breast of the puffin is so distinctive that in one Inuit language they are called *katukh-puk*, meaning "big white breast."

AGAINST ALL ODDS

Young puffins, like those taken by the National Audubon Society to reestablish a colony, face many dangers in such circumstances. While small, they are very prone to being preyed upon by gulls, for instance. But against all odds, Project Puffin proved a positive exercise. Similiar puffin restoration experiments have also taken place in other regions of Maine, and these have had equally good results.

The scientific name for the Atlantic puffin is *Fratercula arctica*, meaning "northern little brother." But they have also been nicknamed "parrots of the sea" because of their very brightly coloured beaks.

SCIENTIFIC NAME: *Fratercula arctica*
say: FRAT-ERK-OO-LAH ARK-TIK-AH
HABITAT: north and middle Atlantic islands
DATA: sea bird; nests in colonies; eggs are incubated by both parents; principal problems leading to decline include oil spillages, overhunting, and reduced availability of fish, an important part of their diet; status: endangered

Rare birdwing

Found only in the forests of Papua New Guinea, the Queen Alexandra birdwing butterfly is now extremely rare. The eruption of a local volcano destroyed much of its habitat back in 1951, and since then it has been further threatened by the continual clearing of areas for human settlement and farming. The use of pesticides has not helped either.

To the *left*, you can see an illustration of the markings of the male of the Queen Alexandra's birdwing; and to the *right* is the female The caterpillars of both sexes, meanwhile, are reddish black with red pointed tentacles. There is also a curious cream "saddle" in the centre of the larva of these beautiful insects. In fact, it is precisely because they are so exquisite that collectors are prepared to pay high prices for specimens; and this, in turn, has become one of the prime factors leading to their demise in the wild. It is also one of the reasons why the government of Papua New Guinea set up a programme to protect the species. But financial help has also helped establish butterfly farms to "feed" the demand for specimens.

SCIENTIFIC NAME: *Ornithoptera alexandrae*
say: ORN-EE-THOP-TERRA AL-EX-AND-REYE
HABITAT: Papua New Guinea
DATA: world's biggest butterfly; now extremely rare; even though many eggs laid, few caterpillars survive to become butterflies as they have many predators; other reasons for decline include destruction of natural habitat; status: endangered

Worldwide, road-building and the destruction of natural habitats, as well as excessive use of pesticides, also continue to lead to the demise of many other types of butterflies.

During the course of a lifetime, the female Queen Alexandra's birdwing may lay well over 200 eggs; but some will be attacked or taken by predators. The caterpillars may also be eaten by snakes, lizards, and other creatures.

In comparison with the number of eggs originally laid, few of this species will finally develop to become mature adults. After that, they will have just a short lifespan, and only live for about another four months.

FACTFILE

The Queen Alexandra's birdwing is the world's biggest butterfly. Females of this species have a larger wingspan than the males.

Indian pangolin

The Indian pangolin mother shown in this illustration is wary. She is protecting the burrow that she built in the ground with amazing speed, using only her clawed front limbs. She shares this home with her mate and the baby pangolin to which she has recently given birth. The male is larger and has strong paternal instincts, taking on care of the young, too, though it is the mother that usually carries her baby about. Normally, pangolins only come out at night; but this one is out by day because she has been dislodged from her home by a hunter. What could he be after? And why has he bothered to disturb this shy and timid creature? All is about to be revealed.

SCIENTIFIC NAME: *Manis crassicaudata*
say: MAN-IS KRAS-EE-COW-DAH-TAH
HABITAT: forests in India, Nepal, Sri Lanka
DATA: formerly hunted for its flesh and because thought to have magical powers; also known as scaly ant-eaters; feed on ants and termites; long tongues and strong claws; nocturnal mainly; status: endangered

The Indian pangolin is a most unusual and increasingly scarce mammal, now listed as severely endangered. Is this any wonder when some of the tribal peoples that share the pangolin's natural habitat in the forests of Nepal, India, and Sri Lanka find its flesh very delicious?

They also sometimes use the creatures scales to make charms said to provide magical cures. Some of these hill tribes, meanwhile, believe pangolin meat is an aphrodisiac.

These nocturnal creatures are easily caught when hunters dig into their burrows. But they do try to defend themselves by rolling up into a tight ball. If this is not effective against a predator, they will resort to squirting their urine!

FOLK CURES

The strong scales of the Indian pangolin, for which it is hunted in addition to its flesh, are basically hair and similar to rhino horn in their structure. There are 28 rows of such scales. Like rhino horn, too, they feature in local folk medicine, and are thought to be able to cure both internal bleeding and rheumatic fever.

Pangolins are also known as scaly ant-eaters. True to this alternative name, they feed on ants and termites, too. Sometimes they will even climb a tree to reach an ants' nest, and then tear it to bits to get at their next meal.

Alternatively, they will break into a termite mound with their strong claws and then introduce their extremely long tongues. These are only withdrawn when lots of termites become stuck so that they can feast greedily.

SILLY SCARE

The Indian pangolin is in fact now so rarely seen that, in October 2000, Indian Express Newspapers reported that a watchman who sighted one felt sure it must be a sort of dinosaur! For some peculiar reason, he linked the small creature's appearance with what he had seen in the movie *Jurassic Park*! It had crept under a truck, and the man even screamed for help. Soon a crowd gathered and, curious, they began to poke at the pangolin with sticks. Eventually, it was taken away by officials and released into a forest.

World of the pangolin

- Pangolins give out a musky smell which helps the males and the females to recognise each other prior to mating.
- Other types of pangolins include the Chinese pangolin, the Malay pangolin, the giant pangolin of Africa, and the long-tailed pangolin.

Slow loris

SCIENTIFIC NAME: *Nucticebus coucang*
say: NUK-TEE-SAY-BOOS KOO-KANG
HABITAT: rain forest in South-East Asia
DATA: lives in trees; gives off smell to ward off predators; hunted in the past for its fur; destruction of habitat also led to decline; killed by some tribal people as believed that their fur had healing powers; status: endangered

The slow loris's greatest enemies are human beings, in the form of poachers, and also large members of the cat family and snakes. It sleeps curled up during the day and so is very vulnerable while it is light, even though probably well camouflaged in the treetops where it lives.

A solitary creature and about the size of a small cat, the slow loris lives in the rainforests of South-East Asia and Indonesia, and is entirely arboreal. It has thick woolly fur, and its colouring varies from a brownish shade to grey, with a light facial streak, as you can see in the illustration, shown *left*.

On each front limb there is a long curved digit which is raised and very useful for scratching. Its hind limbs, meanwhile, are so strong that it can hang upside down to feed on fruit, insects, eggs, and small birds.

Their name may indicate that they are far from speedy in their movements; but if they are disturbed or looking for food, they become very energetic.

Slow lorises have a clever way to mark a particular tree as their own territory, and do this with urine. The urine is spread over the branches and the scent of one slow loris's urine then keeps others away.

No one is sure how many survive in the wild or in protected areas – there are perhaps one million; but we do know their numbers have been in decline for a number of years.

FEWER AND FEWER

How, then, have slow lorises come to disappear so dramatically? They are so cute that some are collected and sold as pets, though this probably distresses them and it is doubtful they can survive for long outside their natural environment. Others have been lost due to destruction of their habitat for agricultural land. There are tribal people, too, who believe the fur of the slow loris can help to heal a wound. The fur has also been used by the fashion industry. How cruel it is, too, that their eyes are sometimes taken to be sold as lucky mascots!

World of the slow loris

- This creature has an intriguing way of warding off predators. It gives off a very smelly substance if threatened, and this foul aroma warns its attacker that it will experience a very unpleasant taste, if it dares to take a bite out of the slow loris.

Mandrill

The male mandrill is among the most spectacular of all monkeys with its very colourful face and hind quarters. It is, however, now among the rarest monkeys in the world due to hunting and trapping.

Found in the rainforests of West Central Africa, the mandrill has a bright red nose, and a blue and red rump, this vibrantly coloured posterior perhaps acting as a visual signal to its young when on the move. One male may live with up to 10 females, reproducing with them all so that in a single group there can be several offspring. Sometimes such groups will also join together and move around in large communities.

The females are smaller than the males and usual confine themselves to the trees. The males, however, often come down to the ground. All feed on fruit, nuts, seeds, fungi, insects, spiders, worms, and small animals.

World of the mandrill

- Scientists think that the more aggressive and colourful a mandrill is, the more offspring he is likely to have.
- Mandrills have enormous canine teeth which tends to make them look more fearsome than they actually are. In fact, they are often quite shy.

PECULIAR HABIT

One of the strange behavioural traits of the mandrill is the way in which the males particularly will find a tree and rub themselves against it so that it takes on their scent. The reason for this peculiar habit remains a mystery to scientists, but it is as if they are somehow marking territory or perhaps demonstrating their sexuality and importance.

The total length of an adult male mandrill is about 36ins (92ms), and a fair part of this measurement is its head, which is very large in comparison with its body size. You could expect the fully-grown male mandrill to weigh about 55 pounds (25kgs), but a female would weigh less than half of this.

In general, very little is known about the mandrill's way of life in the wild; and now that they have become so scarce, it is unlikely that much more will ever be discovered. They move so quickly that it is hard to keep track of them in the forest environment, and they are timid by nature. The only way to observe them for long periods would be to build a hide of some kind which they might gradually come to ignore and accept.

SCIENTIFIC NAME: *Mandrillus sphynx*
say: MAN-DRIL-OOS SFINKS
HABITAT: rainforests of West Central Africa
DATA: among rarest monkeys; decline due to hunting; destruction of rainforest habitat also to blame; females are arboreal; males, terrestrial; diet includes fruit, fungi, insects, nuts, roots, seeds, and small vertebrates; status: endangered

Bald eagle

The only eagle that is unique to North America is the bald eagle. In fact, it is the national bird of the United States, even though most are found in Alaska. You might well ask why it is called "bald" when, clearly, it has a great many black and white feathers. The answer is that "bald" has not always meant just "without hair" but was also used as an alternative word for "white."

Three feet (almost 1m) from head to tail and with a wingspan measuring as much as 6ft (nearly 2m), the bald eagle has long been the pride of America. In the 18th century, when it was first adopted as the American national symbol, there were very many. An estimate is 75,000. But by 1940, so few remained that the Bald Eagle Protection Act was passed, making it illegal to kill, sell or possess this bird without a special permit. Subsequently, it was also officially declared an endangered species. Steps clearly had to be taken to protect it.

World of the bald eagle

- In the United States, anyone caught with a single feather or any other part of a bald eagle is liable to a large fine. However, those Native Americans who use them in their regalia are excluded from this penalty.

SCIENTIFIC NAME: *Haliaeetus leucocephalus*
say: HAL-EE-EYE-ET-OOS LOO-KOS-EFF-AL-OOS
HABITAT: mostly Alaska in North America
DATA: national bird of United States; "bald" part of name means "white;" Bald Eagle Protection Act prevents taking these birds, their eggs, and nests; status: endangered

FACTFILE

Some bald eagle nests are enormous and, added to annually, may finally measure up to 10ft (about 3m) in diameter.

Many factors had led to what at the time was a tragic decrease in numbers – among them, water pollution, human interference, wholesale killing of the species by farmers and ranchers as this bird regularly preyed on their livestock, and widespread use of the pesticide DDT. But, fortunately, conservationists came up with a way of trying to increase their numbers. First, they phased out use of lead shot by waterfowl hunters, so that predatory bald eagles would not be poisoned when feeding on these.

The conservationists set about removing the first clutch of eggs from many bald eagle's nests since many do not hatch. Then they incubated these eggs artifically, later returning the chicks safely to the wild. Other adult pairs of bald eagles seem very willing to foster these "orphans."

Another method of introducing the birds to areas where they had become rare was to keep them in enclosed artifical nests from eight weeks of age, feed them there, and then release them when they were ready to live independently.

As a result of such efforts, the estimated bald eagle population is now as high as 50,000, so in some respects it may be considered almost fully recovered. Indeed, it is a prime example of how the road to extinction can often be reversed.

If the population further increases, the distinctive white head and tail feathers of the bald eagle may again be a familar sight in the skies over North America.

Glossary

ammonites extinct, soft-bodied aquatic creatures with tentacles and coiled shells
aphrodisiac a substance said to make people feel excited
aquatic of the water
arboreal living in trees
archosaurs reptiles, including crocodiles and dinosaurs
asteroid one of the many small planets that orbit the sun between Mars and Jupiter
baleen describes a whale with horny plates growing from the palate in its mouth
blubber a whale's fat
brachiopod a type of shelled animal
brontotheres prehistoric group of animals, resembling rhinos
camouflage to hide by colour in natural surroundings
carapace a turtle's or tortoise's shell
conservation saving or preserving the environment or a species
coprolites fossilized excrement
Cretaceous an era of time, lasting from 144-65 million years ago
creodonts a group of carnivores living from 55-36 million years ago
crocodilian belonging to the crocodile family
cryptozoology the study of strange animals
Devonian an era of time lasting from 408-360 million years ago
DNA a substancr in animal and plant chromosomes that is responsible for hereditary characteristics
environmentalist someone who looks after and studies the environment
Eocone an era of time around 50-36 million years ago
evolve adapt and change over a period of time
evolution adaptation and change over a period of time
extinct no longer part of life on Earth
extinction disappearance from life on Earth
fangs huge canine teeth
fauna animal life
feline of the cat family, or catlike
flora plant life
fossilized embedded and preserved in rock, resin, or other material
gastroliths stones swallowed to help digestions of tough plants
habitat the environment of a creature or plant
hadrosaur a duck-billed dinosaur group
Jurassic an era, lasting from 213-144 million years ago
lagomorph rabbitlike creatures from Eocene times
mammal an animal that gives birth to and suckles live young
marsupial a mammal with a pouch for its young
migration moving in herds to another place
Miocene an era of time about 25 million years ago
monotreme an animal with a single opening for its reproductive and digestive organs
notoungulates a group ofextinct South American herbivores
Oligocene an era of time about 40 million years ago
ornithologist someone who studies birdlife
palaeontologist a scientist who studies fossilized remains
pelt skin
pelycosaur a sail-backed reptile such as Dimetrodon
perissodactyl scientific name for a creature with either a single hoof or three hoofs on each limb
Permian times an era, prior to Triassic times, at the end of which there was a mass extinction
placoderms a group of armoured fish from Devonian times
Pleistocene an era of time about 1.8 million years ago
protuberance a swelling
pterosaur flying reptile from the time of the dinosaurs
quarry a predator's prey
reserve a special place where animals are protected
sauropods long-necked, plant-eating dinosaurs, mainly from Jurassic times
semi-aquatic living partly in water
terrestrial of the land
Tertiary an era of time lasting from 65-2 million years ago
testudinate a creature with a shell, such as a turtle
Triassic an era of time, lasting from 249--213 million years ago
ungulate a hoofed animal

Index

Index